Selection and Use of Preservative-Treated Wood

Daniel L. Cassens
Professor of Wood Products
Department of Forestry and Natural Resources
Purdue University
West Lafayette, Indiana

William C. Feist
Research Chemist (retired)
USDA Forest Service
Forest Products Laboratory
Madison, Wisconsin

Bruce R. Johnson
Research Forest Products Technologist (retired)
USDA Forest Service
Forest Products Laboratory
Madison, Wisconsin

Rodney C. De Groot
Research Plant Pathologist
USDA Forest Service
Forest Products Laboratory
Madison, Wisconsin

Publication No. 7299

Forest Products Society
2801 Marshall Court
Madison, WI 53705-2295
phone: 608-231-1361
fax: 608-231-2152

Disclaimer

The use of trade or firm names in this publication is for reader information and does not imply endorsement by the U.S. Department of Agriculture of any product or service.

Published 1995 by the Forest Products Society.

ISBN 0-935018-75-1

Printed in the United States of America.

95081000

Caution

This publication reports information on pesticides. It does not contain recommendations for their use nor does it imply that the uses discussed here have been registered. All uses of pesticides must be registered by appropriate State and Federal agencies before they can be recommended. The pesticides-wood preservatives, mildewcides, and fungicides described in this report were registered for the uses described at the time the report was prepared. Registrations of pesticides are under constant review by the Environmental Protection Agency. Therefore, consult a responsible State agency on the current status of any pesticide. Use only pesticides that bear a Federal registration number and carry directions for proper use.

Pesticides used improperly can be injurious to humans, animals, and plants. Follow the directions and heed all precautions on the label. Avoid inhalation of vapors and sprays; wear protective clothing and equipment if these precautions are specified on the label.

If your hands become contaminated with a pesticide, do not eat, drink, or smoke until you have washed. If you swallow a pesticide or if it gets in your eyes, follow the first aid treatment given on the label and get prompt medical attention. If a pesticide gets onto your skin or clothing, immediately remove the clothing and wash skin thoroughly.

Store pesticides in original containers out of reach of children and pets, under lock and key. Ask for a Consumer Information Sheet when purchasing treated wood. Follow recommended practices for disposing surplus pesticides and containers. Preservative-treated wood or finished wood should never be burned, either for heat or disposal. Toxic fumes and ashes may be released. Dispose of the material using procedures approved by your appropriate State agency (usually through ordinary trash collection or burial).

Acknowledgments

Several people were asked to review this publication. The authors are indebted to Mr. William J. Baldwin, Dr. H. Michael Barnes, Dr. Stan Lebow, Mr. Donn E. Keefe, Dr. William S. McNamara, Dr. Jeffrey J. Morrell, Mr. Thomas D. Searles, Mr. David A. Webb, and Dr. Jerry Winandy for responding to our request to review and providing useful expert comments. Portions of the material and photographs used in this publication were excerpted from U.S. Navy Document "Wood Protection." The reference is Wood Protection NAVFAC MO-312, May 1990, Naval Facilities Engineering Command, 200 Stovall Street, Alexandria, Virginia 22332-2300. Wood Protection was prepared by Daniel Cassens, Professor of Wood Products, Purdue University, and Rodney De Groot, Research Plant Pathologist, USDA Forest Service, Forest Products Laboratory, with funding from the U.S. Navy.

Summary

For centuries wood has been one of the most common and easiest to use construction materials. It is used practically everywhere in housing and agricultural buildings, as posts, utility and telephone poles, railroad ties, wharves, piling, and many others. Wood is also a renewable material. Although we have consumed tremendous volumes of wood for centuries, the United States is growing more wood volume than what is being harvested.

Wood is a biological material and subject to decay, insect, and marine borer attack. These agents are nature's way of recycling wood in the natural ecosystem. Without nature's recycling system, we would literally be buried by wood and other cellulose-based materials, such as grass, leaves, and agricultural field residues. However, when wood is used in a more or less permanent application, it must be protected from biological degradation. Destruction can be prevented by any number of methods or combination of methods. This book was written for homeowners, contractors, building supply clerks, architects, and others who use or recommend wood products. This summary briefly reviews the major causes of biological deterioration of wood and how it can be prevented.

Agents of Biological Deterioration

Wood decay, or wood rot, is one major cause of wood deterioration. Decay is caused by certain classes of fungi that use moist or wet wood as a food source. As decay develops, the wood is literally dissolved, rapidly loses its strength, and most commonly becomes brown and crumbly (brown rot); or it may develop into a white, soft, fibrous mass (white rot). Dry wood does not decay because decay fungi require moisture. Generally, the term dry rot is a misnomer in that the wood was actually wet when it decayed and then subsequently dried out. Thus, it is sometimes termed dry rotted. Occasionally, dry wood is decayed by the uncommon water-conducting fungi that are able to move moisture to dry wood from another source.

Decay fungi are spread by tiny spores that are equivalent to seeds. These spores are everywhere in the air and soil around us. There is no avoiding them. When a spore comes in contact with a moist piece of wood, it may germinate and develop microscopic hyphae that can quickly spread through the wood, using it for food. Wood decay fungi must have four basic requirements to destroy wood: a food source or the wood itself, moisture, warm temperatures, and oxygen. Eliminating any of these four requirements will stop the decay fungi. Normally, wood does not decay because it is kept dry as in houses or the wood is made a nonfood source by treating it with a preservative. Similarly, wood buried deep underground or submerged in water undergoes very little decay because of limited oxygen.

Insects represent the second major category of organisms that destroy wood. Some insects use the wood as a food source, and a few use it for shelter. Termites are probably the most common insects that use wood as a food source. The common subterranean termite actually nests in moist areas, usually the soil, and moves to the wood and consumes it as a food source. These termites may build shelter tubes over nonpenetrable surfaces in order to reach the wood and not expose themselves to predators and the drying effect of the air. Numerous other wood-boring insects also use wood as a food source. These include other types of termites, beetles, and wasps.

Carpenter ants and carpenter bees are examples of insects that use wood for shelter. Carpenter ants prefer damp and partially decayed wood. They simply gnaw the wood away to create cavities for themselves. When carpenter ants and decay occur together, serious damage can result. Carpenter bees, which look like bumblebees, simply bore into soft-textured wood located under eaves or similarly protected locations and lay eggs in the tunnel created. Damage is normally limited to the ½-in. round entrance holes and tunnels within the wood.

Marine borers represent the third general class of organisms that can destroy wood. There are many different types of marine borers. Their natural ranges, the manner in which they destroy wood, and control measures all vary greatly. Marine borers, regardless of type, are serious pests if present in significant numbers. Wood used in all U.S. coastal and brackish waters is subject to at least some marine borer damage. The number of different species of marine borers and the severity of attack are more pronounced in warmer waters. Some marine borers are tolerant of certain wood preservatives.

One group of marine borers, generally called shipworms, is probably the most fascinating. The young larvae enter the wood by boring a hole less than 1/16 in. in diameter. After metamorphosis, the shipworms continue to grow and bore in the wood. They can never leave it, but their offspring can. One group of shipworms can enlarge to about a ⅞-in. diameter and grow to 5 or 6 ft in length.

Gribbles represent another common group of marine borers. Gribbles are small crustaceans that bore holes about 1/16 in. in diameter and seldom go more than about ½ in. below the wood surface. However, there may be 300 to 400 organisms per square inch. As the wood is honeycombed, it is broken away by mechanical wave action, and the borers continue to destroy the wood, usually most rapidly in the intertidal zone. In piling, a typical hourglass shape results.

Control of Wood-Destroying Organisms

Wood-destroying organisms or their effects can be controlled by a number of different methods or combination of methods. These methods include construction techniques to keep wood dry, insecticides, and the use of naturally durable wood or preservative-treated wood.

When wood can be kept dry to prevent decay and away from the soil to prevent subterranean termite attack, the chances for deterioration are greatly reduced. Traditional house designs are partially the result of designers' and builders' wishes to keep wood dry and off the ground. Masonry materials are generally used for the foundation, and the lowest wood members are usually at least 8 inches from the soil line. This technique keeps the wood from soaking up water from the soil and forces termites to build shelter tubes (thus, announcing their presence) in order to reach the wood above. Wide roof overhangs and gutters help divert water away from the house. Traditional siding and caps over windows and doors are constructed in an overlapping fashion so any wind-driven water is shed downwards. Adequate ventilation in attics and crawl spaces is also provided. When attention is paid to design features and construction detailing that keeps wood dry, centuries of good performance can be expected.

There are many uses for wood where it cannot be kept dry. These include applications such as decks, fence posts and boards, utility and telephone poles, railroad ties, landscape timbers, and marine structures. In these cases, wood treated with preservatives or naturally durable wood species should be specified.

Before the science of wood preservation was developed, woods with naturally durable heartwood were commonly available and used. These woods include black locust, white oak, baldcypress, and cedar in the eastern United States and redwood and most cedars in the western part. The heartwood, or dark central part of these trees compared with the white outer sapwood, is naturally impregnated with extractives that impart some resistance to decay and insects. The natural resistance varies greatly and is less in the younger trees that are available today. It is probably best not to use this material where a high potential for decay or insect attack exists. Today, it is much more common to use wood materials treated with a preservative.

Wood preservation is an inexact science, largely because of the biological origins of wood and the nature of the organisms that degrade wood. From an industrial aspect, numerous standards and regulations have been prescribed. An understanding of preservatives and preservative-treated wood will facilitate intelligent selecting, specifying, purchasing, and installing proper materials.

Depending on how wood preservatives are applied, several different levels of protection can be achieved. Treatments such as brush, spray, and dip are superficial and do not result in deep penetration into the wood, nor can the wood normally absorb large amounts of preservative. Therefore, these treatments offer minimal protection and are generally used for above ground applications where a high decay hazard does not exist, for remedial treatment, or field treatment where treated wood has been cut.

Commercial pressure-treating processes use various combinations of vacuum and pressure to assure that adequate preservative quantities are forced deep into the wood. To ensure long-term performance, the correct preservative for a particular application must be specified. An adequate amount of preservative must then be impregnated into the wood. This is called retention; minimum retention levels are usually specified in treating standards. In addition, the treatment should penetrate the wood as deeply as can be reasonably achieved. Penetration levels are also specified in treating standards.

Essentially, the treating process produces a protective shell of wood-containing preservative around an inner core of untreated material. This shell must not be broken by sawing, boring, machining, or from checks due to seasoning, because the core of untreated wood will then be subject to attack by biological agents. Large wood members will have substantially more untreated core than small members for any given species. If the shell is broken, the exposed wood should be liberally treated by brushing or soaking with a preservative. Smaller dimension lumber and plywood, when properly treated to existing standards, should not present a serious problem.

For wood products, such as lumber, plywood, and posts available in retail yards, waterborne preservatives are most commonly used. Water is used as a carrier for the solution that is forced into the wood. The water subsequently moves out of the wood, leaving the preservative behind. The preservative changes its chemical form and becomes fixed in the wood and highly leach resistant. These preservatives are also referred to as CCA (chromated copper arsenate), which is used commonly in the eastern and central United States on Southern Pine, and ACZA (ammoniacal copper zinc arsenate), which is used on Douglas-fir in the western part. Ammoniacal copper quaternary compound (ACQ), a newer waterborne preservative, was recently introduced. Wood treated with waterborne preservatives should have a residue-free surface. Waterborne preservatives are odor free and, when dry, the treated wood can be stained or painted. Wood treated with waterborne preservatives is commonly used wherever resistance to decay and insects is needed. These uses include decks, sill plates, outdoor steps, landscape timbers, posts and fence boards, posts in pole barns, wharves, and many other applications. If moisture is present when standard ferrous (iron) fasteners are used with waterborne preservatives, corrosion can occur. Therefore, hot-dipped galvanized fasteners are used above ground and stainless steel fasteners are used in severe applications such as soil contact.

Pentachlorophenol (penta) and creosote are probably the most common of the older, traditional, oil-based preservatives, and most wood treated with these preservatives ends up in commercial applications. For example, pentachlorophenol is commonly specified for treating poles, whereas creosote is specified for treating railroad ties and marine structures. Because these preservatives are oil-based and have a characteristic odor, they are generally not suitable for use where skin contact is likely. After aging for many years, however, the surface residue and odor can become less of a problem. Another oil-based preservative called copper naphthenate is also gaining acceptance in the pole market.

As mentioned, numerous standards and regulations exist in the wood treating industry. The standards are developed and maintained by the American Wood Preservers' Association (AWPA). These standards are often revised, and the reader should be sure to obtain the current version. Preservatives are regulated by the U.S. Environmental Protection Agency (EPA), and regulations are often updated and revised. The American Lumber Standard Committee (ALSC) is the organization that accredits agencies that oversee the labeling of wood products treated according to accepted AWPA standards. Products meeting these standards are marked with ink stamps, tags, or brands indicating the supervising agency, treating company, preservative, and other important information. For preservative-treated lumber obtained at lumber yards, producers will normally supply a limited warranty. Wood products not marked with the noted information may not be treated according to industry standards for the particular end use under consideration.

It is important to recognize that the preservatives CCA, ACZA, pentachlorophenol, and creosote are restricted-use pesticides based on EPA regulations. That is, they are only available to individuals who have received special training concerning the safe application and handling of restricted-use pesticides. However, wood products treated with these preservatives are not restricted and they are available to anyone. The EPA has published Consumer Information Sheets providing information on the safe use, handling, and disposal of the treated material. These sheets are available from the material distributor. Be sure to obtain a copy and read it carefully.

Contents

Chapter 7
Standards, Quality Assurance, and Specifications

Chapter 8
Post-Construction Inspections and Remedial Treatment

Chapter 9
Specific Applications for Treated Wood

Chapter 1
Introduction

Introduction

Durability

Wood, because of its abundance, has always been a common building material in North America. Early settlers quickly learned that each species had unique characteristics that made them particularly suitable for certain uses. For example, the heartwood of black locust, white oak, baldcypress, and cedars was noted for being naturally durable or resistant to decay and insects. As a result, these species and others were used where a high potential for decay or insect attack existed. The practice of using naturally durable woods where needed continued to about the end of the 1800s. At this time, virgin old-growth timber was being depleted in many sections of the country. The industrial revolution and particularly the railroads and utilities created a need for large quantities of timber products that would have a long-life expectancy when used in conditions favoring decay and insect attack. In response, the science of wood preservation was developed to inject preservatives into the more plentiful, nondurable woods to make them more resistant to decay and insect attack.

Timber Availability

It makes good sense to use wood as a construction material. Trees are a renewable resource, and when colonists cut virgin timber and the land was not developed, new stands soon began to grow and mature. Timber can be harvested on a periodic basis much like any agricultural crop. With proper management, this resource can be both utilized and maintained. For example, in 1991 the national net timber growth exceeded the volume harvested by a third, whereas in the 1920s the harvest was double the growth.[22]

Environmental Acceptability

In addition to being renewable, wood is one of the most environmentally acceptable materials. From raw material extraction to finished product, the energy input is 70 times greater for a ton of aluminum than for a ton of lumber, and 17, 3.1, and 3 times greater for steel, brick, and concrete block, respectively, than for wood.[9] (See **Table 1** for SI conversion factors.) Wood is also an excellent insulator when compared to other construction materials. For an equivalent thickness, solid wood is 4 times more efficient as an insulator than cinder block, 6 times more efficient than brick, 15 times more efficient than concrete, and 1,770 times more efficient than aluminum.[3] Finally, healthy, fast-growing forests managed by scientific silvicultural methods for long-term yield absorb carbon dioxide from the atmosphere and provide a net oxygen gain, thus reducing the greenhouse effect.

Timber Conservation

In some respects, the wood available today is different from the old-growth material. Tree diameters are smaller and the amount of heartwood is less. Therefore, when the material is used in a high decay or insect hazard application, in most cases, it should be treated with a wood preservative. The proper use of wood preservatives ensures maximum longevity of the material. Two important benefits result. First, if the wood does not have to be replaced, less timber needs to be cut. Southern Forest Products Association (1990) estimated that treating wood saves 226 million merchantable trees annually.[24] Second, replacement of deteriorated wood is costly in terms of the wood used and the labor required as well as general disruption.

Health Issues

Wood preservatives, like many other chemicals, can create environmental and health hazards when not properly handled and applied. Available wood preservatives are regulated by the U.S. Environmental Protection

Table 1. *SI conversion factors.*

English unit	Conversion factor	SI unit
degree Fahrenheit (°F)	(1.8)(C)+32	°C
foot (ft)	0.3048	meter (m)
inch (in.)	25.4	millimeter (mm)
pound (lb)	0.4535	kilogram (kg)
pound per square inch (lb/in^2) (stress)	6,894	Pascal (Pa)
pound per cubic foot (lb/ft^3) (weight)	16.0	kilogram per cubic meter (kg/m^3)
ton (short, 2,000 lb)	907.1	kilogram (kg)

Agency (EPA). Some preservatives are available "over the counter" to the general consumer, while the sale of others is restricted to certified applicators. The restricted-use applies to the chemical preservative and not to treated wood products. Wood products treated with restricted-use pesticides or preservatives are available to the general public. It is important to read all labels and follow proper use, handling, and disposal recommendations. Preservative-treated wood should be used only where conditions warrant use. Safe use of wood preservatives and treated wood products is addressed in Chapter 11.

This publication first provides background information on wood as a construction material as well as biological deterioration. Then, methods used to protect wood, types of preservatives, treatment processes, and properties of treated wood are addressed. How buyers can specify treated-wood products and be assured of receiving a quality product is addressed. Finally, safe handling and environmental concerns with treated wood products are discussed. Emphasis is placed on the softwood species because they are the most common commercial species treated with preservatives. Some information on hardwoods is also provided when appropriate.

Chapter 2
Wood as a Construction Material

Wood as a Construction Material

Lumber, heavy timbers, poles, piles, and many other wood products are derived from trees. Trees are biological organisms we see and enjoy every day, but we seldom consider the complexity and variability of the wood produced by them. An understanding of some basic characteristics and properties of wood can help in making sound decisions for specifying, treating, and using the many different wood products. In addition, a better understanding of wood characteristics can also help solve problems in current applications.

Softwoods and Hardwoods

Although hardwoods are treated, softwoods are the most common material treated with preservatives in North America. Softwoods are the pines, spruces, larch, true firs, hemlock, redwood, yew, baldcypress, Douglas-fir, and cedars having needles or scale-like leaves and, except for baldcypress and larch, maintain these needles throughout the year. They are also called evergreens. Because softwoods bear scaly cones to produce seeds, they are sometimes called conifers. Softwoods are most commonly used for construction lumber, heavy timbers, poles, and piles where functionality, not appearance, is important.

Softwoods for construction purposes are specified by species groupings.[19] For example, if Southern Pine is specified, the shipment may contain any of the major species, namely loblolly, longleaf, shortleaf, or slash pine. Another grouping called Mixed Southern Pine may contain any of the 10 Southern Pine species. For western species, larch can be mixed in with Douglas-fir and sold as the species grouping Douglas-Fir-Larch or Douglas-Fir-Larch (North). Hem-Fir is another mixed grouping composed of western hemlock and several species of true firs. These species groupings are made on strength characteristics and the availability of the species within a producing region.

Hardwoods are trees such as the oaks, maples, hickories, gums, and others having broad deciduous leaves. Their leaves change color in the fall and drop to the ground at least in temperate climates. Hardwoods are often used for railroad ties, heavy timbers, pallets, and interior applications. Hardwoods used for crossties are always treated, but other than this application, the availability of preservative-treated hardwoods is limited.

The terms hardwood and softwood do not directly describe the hardness or softness of the wood. Instead, they refer to the leaf form or mode of seed production for trees from which the wood is cut. The terms are especially confusing because some true hardwoods, such as basswood, have softer or lighter wood than common softwoods such as the southern pines. Softwoods are also very strong for their weight, compared to hardwoods.

Sapwood and Heartwood

The wood formed immediately inside the bark of a tree is called sapwood. Sapwood is light in color and contains living cells that transport water from the roots to the branches and leaves at the top of the tree. Heartwood is formed in the central part of the tree stem, as those water-conducting cells in the sapwood die. As a result of an accumulation of various materials, extractives, pitch, oil, and other extraneous substances, the heartwood is often darker than the sapwood. These materials impart natural decay resistance to the heartwood of some species.

The heartwood of most tree species is difficult to treat with preservatives, but the sapwood of nearly all species accepts preservatives. Southern Pine poles and lumber readily accept preservatives because these are mostly sapwood (**Fig. 2-1**). In contrast, poles and lumber cut from mature Douglas-fir trees contain a high proportion of impermeable heartwood that is difficult to treat. Considering the hardwoods, the heartwood of red oak is permeable, whereas that of white oak is some of the most

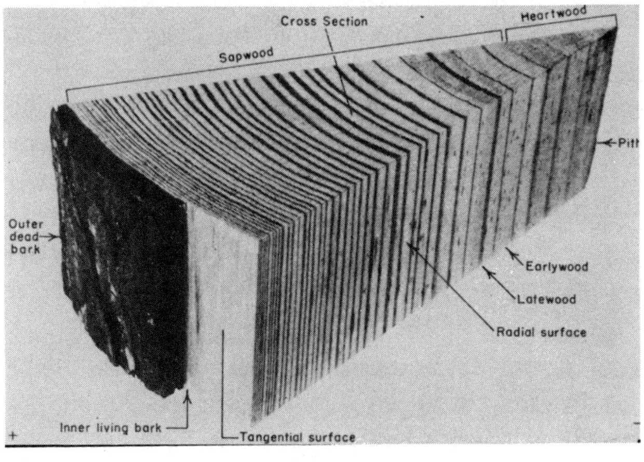

Figure 2-1. *Section of mature pine tree showing gross wood structure.*

impermeable wood. Thus, red oak is preferred for railroad ties. It is this type of variability that shapes the grouping of wood products by species into standards for wood preservative treatments.

Few, if any, differences exist in the mechanical properties of sapwood and heartwood. Heartwood of some species may have a slightly higher weight per unit volume than sapwood because of the presence of significant amounts of extractives.

Natural Durability

In the past, old-growth naturally durable woods, such as cedars, baldcypress, redwood, chestnut, were available and commonly used where biological deterioration could occur. **Table 2-1** ranks the more common species for natural heartwood durability, and **Table 2-2** rates several wood species commonly imported into the United States. It is the extractives, pitch, oil, and other extraneous substances deposited in the heartwood that impart resistance to decay and insects, thus making certain species more durable than others. When the extractives lack toxic or repellent effects, the heartwood is no more resistant than the sapwood. Today, the demand for naturally durable woods exceeds supply, so they are usually too expensive for general construction purposes. Furthermore, second-growth heartwood may be less durable than the scarce old-growth material. As a result, wood properly treated with preservatives is more commonly used in areas of high decay risk than are the naturally durable species.

Earlywood and Latewood

Growth increments in softwood species, such as southern pines and Douglas-fir, are composed of two distinct parts. Earlywood (sometimes called springwood) is that portion of the growth increment formed during the first part of the growing season or usually in the spring. Earlywood cells are relatively large in diameter, thin-walled, and light colored. Latewood (sometimes called summerwood) is formed after the earlywood (**Fig. 2-1**). Latewood cells are small, thick-walled, and dark. As a result, latewood is more dense and stronger than earlywood.

The growth rate or width of the growth increment and the percentage of latewood it contains are important factors in the grading of Southern Pine lumber where mechanical properties are of concern. For example, a grade called Dense Southern Pine is defined as averaging, one end or the other, not less than six annual rings per inch, and a third or more of each annual ring is latewood. Pieces averaging not less than four annual rings per inch with one-half latewood are also classified as dense. Fast-growth southern pines may have only two or three growth increments per inch and a relatively small percentage of latewood in each ring. The mechanical properties of the fast-growth southern pines are less than those for the material with a high percentage of latewood. As a result, the fast-growth material is placed in a lower grade for lumber used in structural applications.

Table 2-1. *Natural heartwood decay resistance of several domestic species.*[12]

Resistant or very resistant	Moderately resistant	Slightly or nonresistant
Baldcypress (old growth)	Baldcypress (young growth)	Alder
Catalpa	Douglas-fir	Ashes
Cedars	Honeylocust	Aspens
Cherry, black	Larch, western	Basswood
Chestnut	Oak, swamp chestnut	Beech
Cypress, Arizona	Pine, eastern white	Birches
Junipers	Redwood (second growth)	Buckeye
Locust, black[a]	Southern Pine:	Butternut
Mesquite	Longleaf	Cottonwood
Mulberry, red[a]	Slash	Elms
Oak:	Tamarack	Hackberry
Bur		Hemlocks
Chestnut		Hickories
Gambel		Magnolia
Oregon white		Maples
Post		Oak (red and black species)
White		Pines (other than longleaf, slash, and eastern white)
Osage-orange[a]		Poplars
Redwood (old growth)		Spruces
Sassafras		Sweetgum
Walnut, black		True firs (eastern and western)
Yew, Pacific[a]		Willows
		Yellow-poplar

[a] Heartwood has exceptionally high decay resistance.

The latewood in softwoods is more easily penetrated with wood preservatives than earlywood, but the retention is less due to a decreased void volume. Wide bands of latewood also make species, such as southern pines and Douglas-fir more prone to peeling and flaking when painted.

Grain Patterns (Planes or Surfaces)

Visible characteristics, shrinkage, and mechanical properties of wood are defined in terms of the three planes or surfaces in which wood can be cut from a log (**Fig. 2-1**). In reality, however, few boards are perfectly cut on one of these planes. The saw cut usually ends up at some angle between the radial and tangential planes.

Table 2-2. *Natural heartwood decay resistance of several imported species.*[12]

Resistant or very resistant	Moderately resistant	Slightly or nonresistant
Angelique	Andiroba[a]	Balsa
Apamate	Apitong[a]	Banak
Brazilian rosewood	Avodire	Cativo
Caribbean pine	Capirona	Ceiba
Courbaril	European walnut	Jelutong
Encino	Gola	Limba
Goncalo alves	Khaya	Lupuna
Greenheart	Laurel	Mahogany,
Guijo	Mahogany,	Philippine:
Iroko	Philippine:	Mayapis
Jarrah	Almon	White lauan
Kapur	Bagtikan	Obeche
Karri	Red lauan	Parana pine
Kokrodua	Tanguile	Ramin
(Afrormosia)	Ocote pine	Sande
Lapacho	Palosapis	Virola
Lignumvitae	Sapele	
Mahogany,		
American		
Meranti[a]		
Peroba de campos		
Primavera		
Santa Maria		
Spanish-cedar		
Teak		

[a] More than 1 species included; some may vary in resistance from that indicated.

End grain (cross section) is the surface exposed when wood is cut across the width of a log or board. The end grain reveals the annual rings. Because the end grain is porous, it absorbs or looses moisture much more easily than the side grain (radial or tangential). The end grain is also more easily treated with preservatives.

The radial surface is exposed when a log is cut longitudinally from its center or pith to the bark (along the radius). In the hardwood industry, lumber cut this way is known as quartersawn lumber. Some softwood species, such as southern pines and Douglas-fir, are quartered to expose a vertical grain pattern and used for flooring because of its high resistance to wear and tendency not to cup. It is also the preferred material for preservative-treated Southern Pine shakes and decking. This lumber is sometimes referred to as edge grained or vertical grained (**Fig. 2-2**).

The tangential surface is exposed when a board is cut parallel to the bark and tangent to the log diameter. The lumber is called flat grained (softwoods) or plainsawn (hardwoods). This is the most common way of producing lumber today. A characteristic U- or V-shaped grain pattern (**Fig. 2-2**) results in the softwoods with distinct earlywood and latewood bands, such as southern pines and Douglas-fir, and the coarse grained hardwoods, such as oak.

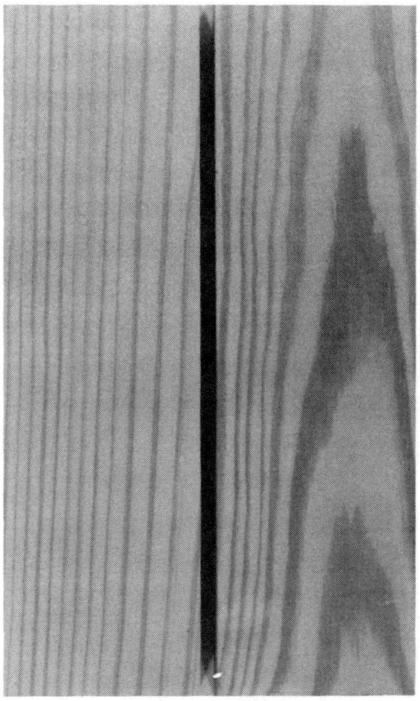

Figure 2-2. *Earlywood (light colored) and latewood (dark colored) bands in both edge (left) and flat grain (right) samples of southern pines.*

Moisture Content Levels

Because the amount of water in wood (moisture content) affects wood treatability, strength, durability, and dimensional stability, wood moisture content is often referenced in specifications.

The moisture content of wood is the ratio of the weight of water in the wood to the weight of the wood when it is ovendry. Wood is considered ovendry when it is heated to 103 ±2°C and approximately constant weight is achieved. The following formula is used to calculate the moisture content of wood:

$$\text{Moisture content (\%)} = \frac{\text{Initial weight} - \text{Ovendry weight}}{\text{Ovendry weight}} \times 100$$

Wood cut from a green log often contains as much or more than its ovendry weight in the form of sap or water, hence its moisture content may be more than 100 percent. The cell walls in green wood are fully saturated and swollen, and the cell cavities are partially to completely filled with water. The water in wood cell cavities is called free water.

After the wood has dried to about 30 percent moisture content, it is at the fiber saturation point. In this state, the cell cavities are emptied of free water, but the cell walls are still saturated with bound water and in their weakest condition. At moisture content levels above the fiber saturation point, wood can be attacked by decay fungi. The expected moisture content of wood in properly constructed and maintained buildings, by region of the country, is given in **Table 2-4**. *Note: Because the moisture content of wood is variable and changes with environmental conditions, wood with a moisture content of 20 percent or more is considered subject to decay.*

Because wood is a hygroscopic material, the amount of bound water the wood holds depends on the relative humidity. That is, wood responds to changes in atmospheric humidity and looses water as relative humidity decreases, regaining water as the relative humidity increases. For a given relative humidity level, a balance is eventually reached where the wood is no longer gaining or loosing moisture. When this balance of moisture exchange is established, the amount of bound water eventually contained in a piece of wood is called the equilibrium moisture content.

Table 2-3. *Shrinkage values from green to ovendry moisture content for selected domestic species.*[12]

	Shrinkage from green to ovendry moisture content (%)[a]	
	Radial	Tangential
Hardwoods		
Oak (northern red)	4.0	8.6
Oak (white)	5.6	10.5
Sweetgum	5.3	10.2
Yellow-poplar	4.6	8.2
Softwoods		
Douglas-fir[b]		
Coast type	4.8	7.6
Interior north	3.8	6.9
Interior west	4.8	7.5
Fir		
Pacific silver	4.4	9.2
White	3.3	7.0
Pine		
Eastern white	2.1	6.1
Southern[c]	5.0	7.6
Western hemlock	4.2	7.8

[a] Expressed as a percentage of the green dimension.
[b] Coast Douglas-fir is defined as Douglas-fir growing in Oregon and Washington west of the summit of the Cascade Mountains. Interior west includes California and all counties in Oregon and Washington east of but adjacent to the Cascade summit. Interior north includes the remainder of Oregon and Washington and Idaho, Montana, and Wyoming.
[c] Average of four major species.

Table 2-4. *Recommended moisture content values for wood at time of installation.*[12]

	Moisture content (average range) (%)		
	Most areas of United States	Dry southwestern areas	Damp, warm coastal areas
Interior Woodwork, flooring, furniture, wood trim, laminated timbers, coldpress plywood	8 (6-10)	6 (4-9)	11 (8-13)
Exterior Siding, wood trim, framing, sheathing, laminated timbers	12 (9-14)	9 (7-12)	12 (9-14)

Shrinking and Swelling

Wood shrinks and swells as a result of the loss or gain of bound water from the cell walls. The amount of movement depends on the amount of water gained or lost, the orientation of the wood cells, and the species (**Table 2-3**). The percentage of shrinkage from green to any given moisture content can be calculated by dividing the total shrinkage provided in **Table 2-3** by 30 and multiplying by the moisture content at less than 30 percent. This movement can be significant and create structural and aesthetic problems.

To minimize problems of shrinking, swelling, checking, and warping, wood should be dried to its anticipated service equilibrium moisture content before installation. **Table 2-4** shows the recommended moisture content values for various wood items at the time of installation.

Shrinkage of normal wood in the longitudinal direction is about 0.1 percent and considered insignificant. Tangential shrinkage in wood is generally 1.5 to 2 times that of the radial shrinkage. Because radial and tangential shrinkage is not equal and most boards contain some combination of the two grain patterns, warping will result if green lumber is allowed to dry without restraint. Warping includes cup, bow, twist, diamonding, and crook (**Fig. 2-3**). Round products such as poles, piles, posts, and large sawn timbers develop surface checks and deep cracks for the same reason (**Figs. 2-4** and **2-5**). Checks or cracks that penetrate the preservative shell on treated products provide entryways for decay fungi and insects. These products are ideally dried before treating so the preservative will penetrate any openings that develop. In many commercial processes using creosote or pentachlorophenol, poles, railroad ties, and large timbers are put in treating cylinders and steamed or Boultonized (heated in oil) prior to treatment. This process dries the wood to acceptable limits.

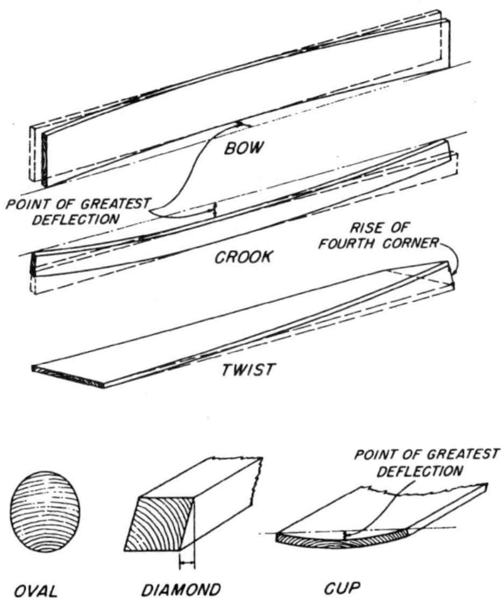

Figure 2-3. *Various types of warp.*

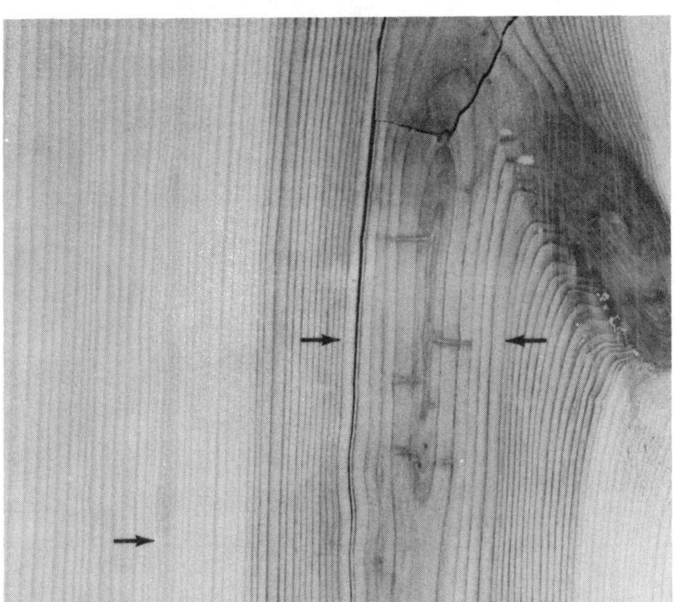

Figure 2-4. *Juvenile wood with cross grain checking (top arrows) and compression wood (bottom arrow) in ponderosa pine.*

Figure 2-5. *Shelling from compression wood in a pine utility pole.*

Treatability

Treatability describes the ease that preservatives can be forced into the wood under pressure. With the possible exception of hemlock and true fir, the sapwood of all species is easily treated. However, as with the many other wood characteristics, heartwood treatability varies with the wood species (**Table 2-5**). Significant variability may also occur within a given species. Douglas-fir heartwood is a good example. The heartwood of coastal Douglas-fir is relatively permeable; Cascade Mountain sources are moderately impermeable; intermountain sources are impermeable or refractory.

Lastly, the treatability of wood varies with grain direction. The penetration ratio of end to side grain for the preservative oils is approximately 15 to 1 and 20 to 1 for the waterborne preservatives.[14]

Table 2-5. *Classification of American woods with respect to heartwood penetrability.*[17]

Softwoods	Hardwoods
Group 1. Easily penetrated	
Bristlecone pine	Basswood
Pinon pine	Beech (white heartwood)
Ponderosa pine[a]	Black gum
	Green ash
	Pin cherry
	River birch
	Red oaks
	Slippery elm
	Sweet birch
	Tupelo gum
	White ash
Group 2. Moderately difficult to penetrate	
Douglas-fir (Pacific-coast type)	Black willow
	Chestnut oak
Jack pine	Cottonwood
Loblolly pine[a]	Largetooth aspen
Longleaf pine[a]	Mockernut hickory
Norway pine	Silver maple
Shortleaf pine[a]	Sugar maple
Western hemlock	Yellow birch
Group 3. Difficult to penetrate	
Eastern hemlock	Hackberry
Engelmann spruce	Rock elm
Lowland white fir	Sycamore
Lodgepole pine	
Noble fir	
Sitka spruce	
Western larch	
White fir	
White spruce	
Group 4. Very difficult to penetrate	
Alpine fir	Beech (red heartwood)
Corkbark fir	Black locust
Douglas-fir (intermountain type)	Chestnut
	Redgum
Northern white cedar	White oaks (except chestnut oak)
Tamarack	
Western redcedar	

[a] Recent experience indicates that the heartwood is variable in penetrability, and these species may be more difficult to penetrate than indicated.

Mechanical Properties

Mechanical or strength properties refer to those characteristics architects or engineers use in the design of a structure (utility line, wharf, bridge, house). These properties, called allowable design values for lumber, are expressed in pounds per square inch (lb/in^2).[19] A complicating factor occurs with wood because it has different strength values across the grain compared to along the grain. Most species are four to five times as strong in compression along the grain as they are across the grain. Wood that is relatively strong with respect to one strength property may be weaker in a different property when compared to another species.

The following are some important wood mechanical properties:

- The modulus of elasticity (MOE or E), a measure of stiffness or rigidity, is an important property for designing wood members that do not sag.
- The modulus of rupture (MOR), a measure of the ultimate ability of a beam to support an applied load, represents the point where a beam will fail. This is considered a life-endangering property, and allowable design values are substantially reduced from published values to provide a safety margin.
- Compression strength, a measure of the resistance of wood to being crushed, is important where heavy loads are involved.
- Tensile strength (basically the opposite of compression strength) is a measure of resistance to

wood being broken by pulling. In this case, the load is applied in such a way that the wood member is pulled from end to end (tension parallel to the grain). An example is the bottom chord or member of a single roof truss.

- Shear strength parallel to the grain, a measure of the ability of wood to resist the slipping of one part along another, is important in beam design.
- Side hardness, a measure of the ability of wood to resist denting, is important in areas such as flooring.
- Toughness, a measure of the amount of work expended in breaking a small specimen by impact bending, is important when impact loads are applied. Toughness is reduced by the early stages of decay and even stain fungi.

Chapter 3
Biological Deterioration of Wood

Biological Deterioration of Wood

Wood preservatives are used to prevent damage by fungi, bacteria, insects, marine borers, and other agents described here. An understanding of these organisms and their damage will contribute to improved wood procurement, construction, or rehabilitation practices. The various characteristics are summarized for fungi and bacteria in **Table 3-1**, for termites and carpenter ants in **Table 3-2**, for wood boring insects in **Table 3-3**, and for marine borers in **Table 3-4**. The correct identification and effective treatment of these numerous pests can require specialized expertise and experience. Licensed pest control operators (PCOs) are individuals who have this specialized knowledge.

Fungi Causing Wood Rot

Wood decay, commonly called rotting, is the fungal decomposition of woody material. Fungi are single-celled or multicellular filamentous (thread-like) organisms that use wood and other plant materials as food. The thread-like structures, visible only with a microscope, are called hyphae (**Fig. 3-1**). Initially, individual hyphae develop from germinating spores and spread throughout the wood. The hyphae secrete enzymes that attack wood cells and finally cause the wood to disintegrate. In its early or incipient stage, wood decay is difficult to detect even though serious strength loss in the wood may have already occurred. Toughness, or resistance to impact bending, is the mechanical property in wood most sensitive to decay. A decrease of a third to half the normal toughness value of wood may occur even before weight loss is detected.

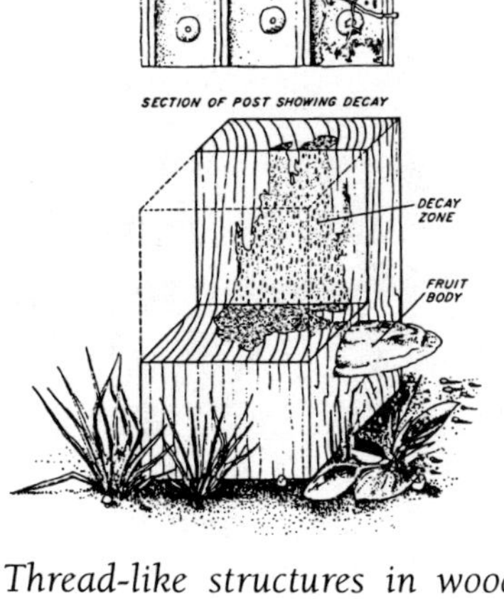

Figure 3-1. *Thread-like structures in wood cells called hyphae.*

Advanced decay is readily detected. Wood with appreciable decay will break brashly (abruptly) across the grain, whereas sound wood splinters at the break. Brashness, reflecting reduced toughness, can be detected by breaking small pieces by hand or lifting pieces of wood from the surface by means of a pointed tool—the pick test (**Fig. 3-2**). The wood should be wet, otherwise it may break brashly even if sound. Brashness is also characteristic of compression wood (discussed in Chapter 2), and test results should be interpreted appropriately.

Fungi Requirements

Fungi have four requirements for growth: food, oxygen, favorable temperature, and moisture.

Fungi are nongreen plants; therefore, they cannot utilize sunlight to synthesize food materials. Fungi require an already synthesized food source, such as wood. Mold fungi generally feed on material located on the wood surface, while sapstain fungi utilize food substances stored in the wood cells. Decay fungi utilize the wood cells or the material holding the cells together (lignin), thus causing serious weight and strength losses. Wood preservatives protect wood by making it nonusable as a food source.

All wood-decaying fungi need a source of oxygen. Wood stored under water or deep in the soil does not readily decay because of the lack of oxygen. Only very slow and superficial degradation by soft-rot fungi and bacteria may occur under these conditions.

The best temperature for most wood decay fungi to develop is 75°F to 90°F. Decay continues at a reduced rate at less than 50°F and slightly greater than 90°F. Decay is dormant near 32°F and 100°F. Fungi may be killed by sterilization treatments. Sterilization temperatures and duration depend on the size of the wood material as well as the fungal species. A typical recommended temperature for eliminating fungi is 155°F for 75 minutes.

For fungi to destroy wood, free water must be present in the cell cavities. Free water is usually found when the wood moisture content is 25 to 30 percent or greater. Thus, one effective way to control wood decay is to keep the moisture content of wood less than 20 percent. Twenty percent is a commonly cited value because wood moisture content changes with environmental conditions. Even within the same wood member, moisture content may vary depending on the immediate surroundings. Lumber properly installed in buildings has a moisture content of 8 to 15 percent, and that cut from freshly harvested trees may have a moisture content of more than 100 percent, based on the ovendry weight of the wood. Wood in soil contact usually has a moisture content of 20 percent or more.

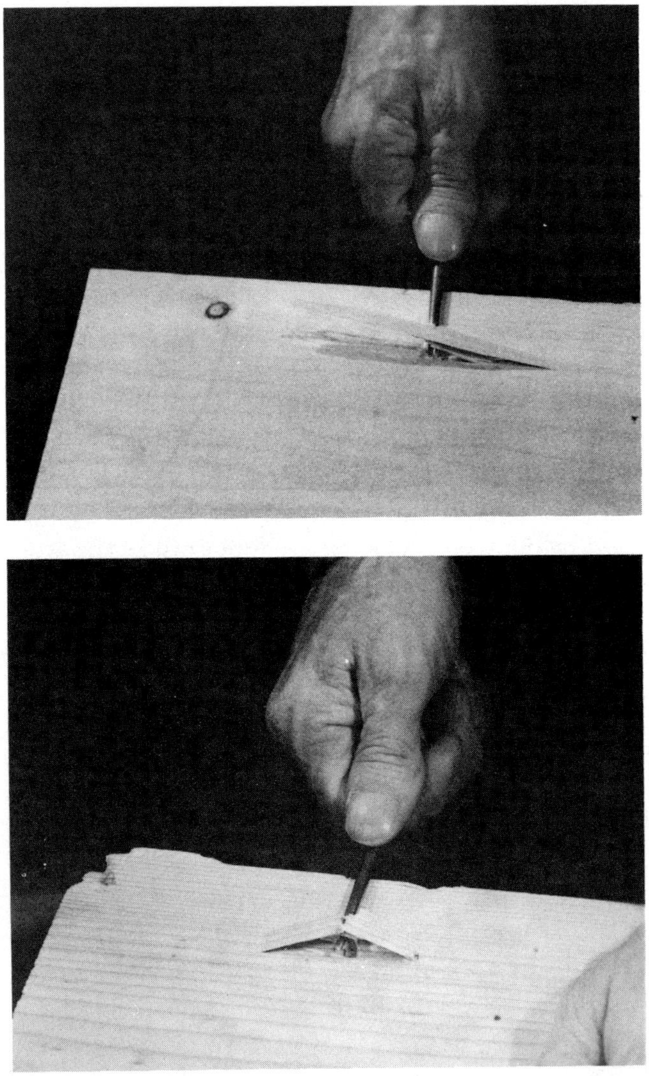

Figure 3-2. *When wet wood is probed with a pick or comparable tool, it tends to lift out as a long sliver or breaks by splintering if sound (top), but if decayed even slightly, it tends to lift in short lengths and to break abruptly without splintering (bottom).*

The amount of moisture available is affected by geographic location. **Figure 3-3** shows the relative decay potential for different parts of the United States.[4] The

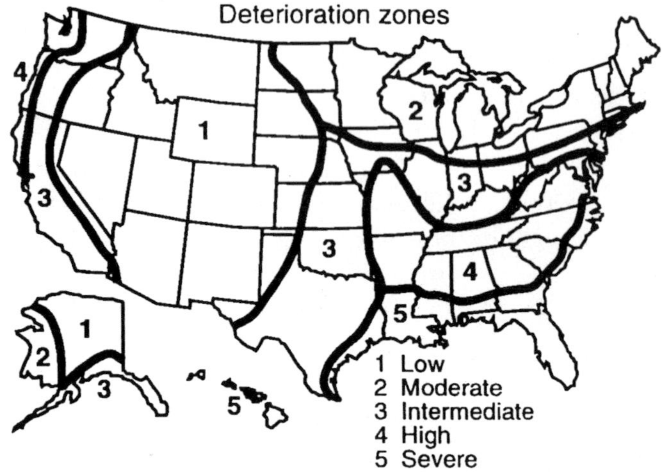

Figure 3-3. *AWPA map showing the relative decay hazard for different regions in the United States.*

Figure 3-4. *Early stage of brown rot, showing discoloration on side grain (top right) and end grain (top left) of the same board and across-grain cracking and collapse (bottom).*

southeastern United States and the Coastal northwest have the highest potential for decay, and the dry southwest has the least.

Brown Rot

Brown rot (**Fig. 3-4**) is the predominant type of decay that can develop in softwoods used above ground. Brown-rot fungi remove the carbohydrates (cellulose and hemicellulose), leaving a modified lignin residue. In the early stages of decay, the wood surface lacks luster and appears dull or dead. The strength properties of wood attacked by brown-rot fungi decrease rapidly even in the early stages of decay. As decay progresses, the wood acquires an abnormal brown color, often appearing as if it has been charred. Cross-grain cracking, collapse or crumbling, and abnormal shrinkage finally result.

The term dry rot is usually used erroneously in describing heavily decomposed, brown-rotted wood and is applied to wood that is both rotten and relatively dry. Actually, in most cases, the wood was wet when it decayed and subsequently dried out. A few fungi can transport water to wood that would otherwise be dry and thereby cause its decay.

Table 3-1. *Comparison of characteristics of brown rot, white rot, molds, sapstains, soft rot, and bacteria-infected wood.*

Characteristics	Brown rot	White rot	Mold	Sapstain	Soft rot	Bacteria
Color	Initially lacks luster and appears dead then abnormal brown color develops	Initially off-white, sometimes with black zone lines	Black, green, orange or other shades	Blue to black, gray or brown, some red, purple or yellow	Darkened, dull brown to blue gray	Brownish, inner zones have been noted to be greenish
Preferred host	Softwoods	Hardwoods	Surface of both sapwood and heartwood for softwoods and hardwoods; sapwood preferred	Sapwood of both softwoods and hardwoods	Softwoods and hardwoods	Sapwood and earlywood more susceptible than heartwood and latewood
Strength	Significant, rapid reduction in many properties	Gradual reduction except for toughness which is rapidly reduced	No serious effect except toughness	Only toughness reduced	Significant in thin members	Significant reductions in many properties with prolonged exposure
Surface	Cross-grain checking, collapse or crumbling and abnormal shrinkage	Nearly normal until advanced stages; white fibrous mass results	Can be brushed or planed from surface; shallow staining may result in hardwoods	Discolored blue to black even when planed	Softened on surface; fine cracking and fissuring both with and across the grain when dry	Softened on surface; cross checking develops when dry
Other			Can occur on air-dried wood when relative humidity exceeds 90 percent		Attacks surface of saturated wood	Attacks water saturated or completely submerged wood

White Rot

White-rot fungi preferentially remove lignin or lignin and cellulose together. White-rot fungi (**Fig. 3-5**) are important in hardwoods used above ground. In the early stages of decay, the wood tends to turn an off-white and sometimes appears bleached. Black zone lines may develop in the light areas. Unless severely degraded, the wood does not crack across the grain and develop abnormal shrinkage or collapse as with the brown rots. A white fibrous mass may occur in the late stages of decay. With the exception of toughness, strength properties decrease gradually.

Molds and Sapstains

Mold and sapstain fungi are two additional important types of fungal damage. Molds (**Fig. 3-6**) discolor the surface of wood by turning it different shades of green, black, and occasionally orange or other light colors. Although the fungal hyphae do penetrate into the wood, the surface discoloration can generally be planed or even brushed off. Strength, other than toughness, is not seriously affected by molds, but absorptivity may be markedly increased. Some molds can develop on airdried wood with a moisture content in the 20 percent range if relative humidity is 90 percent or more.

Sapstain fungi cause a (**Fig. 3-7**) blue to black, gray or brown darkening of the sapwood. The dark color is due to the deep penetration of large masses of fungal hyphae. Some sapstains may produce relatively bright colors such as red, purple, and yellow. In warm weather, the sapwood of some species can discolor in less than a week. It is generally agreed that sapstains do not seriously reduce wood strength except for toughness. Therefore, the wood should not be used where it will receive significant repeated jars, jolts, or blows. However, it should be

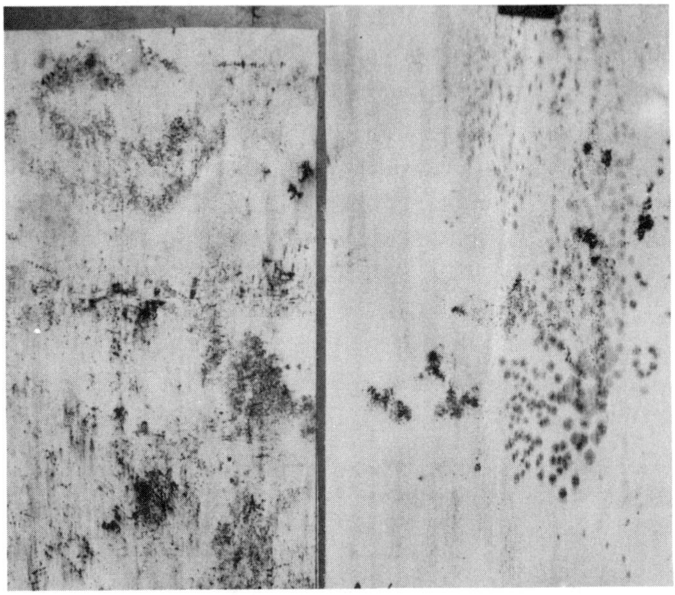

Figure 3-6. *Surface mold on sweetgum veneer.*

Figure 3-5. *Discoloration from white rot on end grain, characterized by mottling and dark zone lines (arrows) bordering the abnormally light-colored areas (top) and side grain of the same board (bottom).*

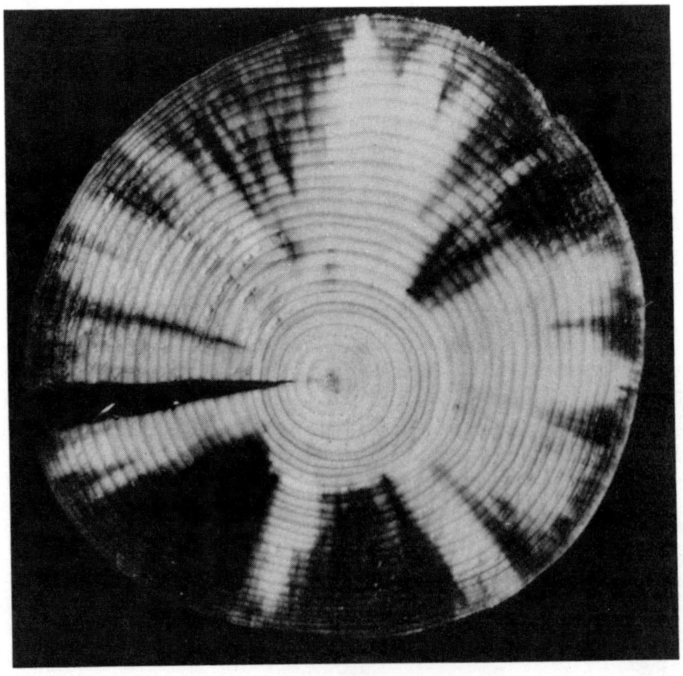

Figure 3-7. *Sapstain in a cross section of pine.*

recognized that decay could easily accompany intense sapstain and affect other strength properties.

Two good methods to control molds and sapstains are to dry the wood product to less than 20 percent or treat it with an approved fungicide.

Soft Rot

Soft-rot fungi (**Fig. 3-8**) affect wood exposed to very wet conditions, such as that used in cooling towers and hardwoods treated with waterborne preservatives. The surface of wood that has soft rot can be scraped off with a fingernail. The wood is darkened, dull-brown to blue-gray. When dry, the surface may appear as though it had been lightly charred, and there will be profuse fine cracking and fissuring, both with and across the grain. When probing with a sharp object, the degraded wood is often comparatively shallow and the transition between it and the underlying firm wood is abrupt.

Soft-rot fungi have been shown as a group to tolerate certain extremes of the environment such as high temperatures, increased alkaline conditions, preservatives, and restricted oxygen levels. Oil-type preservatives may provide increased protection to hardwoods exposed to soft-rot fungi in very wet environments.

Bacteria

Bacteria are minute, one-cell organisms that can slowly degrade untreated or poorly treated wood products, such as piles or wood slats in cooling towers, that are completely saturated or submerged in water for many years. Upon drying, the outer, degraded area of the wood develops a cross-checking that is particularly pronounced on the tangential face. The sapwood of all wood species appears susceptible, the heartwood less so. The earlywood may decompose more rapidly than the latewood.

Insects

Insects are the second major category of biological organisms that can damage and destroy wood products. Because the damage caused by insects is usually more easily observed than the insects, emphasis in this section is placed on identifying damage. Two forms of insects, the immature and the adult form, may be involved in creating the damage. The adult form of the insect is present for only a portion of the year. Immature forms may be observed during other times. Thus, correct identification based on the insect is often difficult. The important characteristics are summarized in **Table 3-2** for termites and carpenter ants and in **Table 3-3** for wood-boring insects.

Wood-destroying insects require three conditions to complete their life cycle: (1) a source from which the infestation can spread, (2) susceptible wood, and (3) suitable conditions of temperature and humidity. If any of these three conditions are eliminated, insect activity ends. Insects are often selective in what they will attack. Moist conditions that support fungal decay in wood often encourage insect infestations.

Figure 3-8. *Typical surface checking of soft-rotted wood when dry.*

Figure 3-9. *Range map showing northern limits for subterranean (A) and drywood (B) termites.*

Subterranean Termites

Subterranean termites refer to both native and imported species having specific requirements for moisture. The insects maintain their nests in the ground or in very close contact with moisture sources. Subterranean termites range throughout much of the United States (**Fig. 3-9**).

Native subterranean termites are small-to-medium-sized insects that live in social groups or colonies. These individuals will be found in different stages of metamorphosis (egg, nymph, or adult) and different casts (reproductives, workers, and soldiers). Both winged and wingless adults occur in any one colony.

A termite colony is initiated with two primary reproductives or swarmers. The swarmers are light tan to black in color with four equal-sized wings, three pairs of legs, one pair of antennae, and one pair of large eyes on the head and are about ⅓ to ½ in. long. In most parts of the country, these primary reproductives are released by a mature colony of termites during daylight hours in the spring and early summer. In the desert southwest and southern California, the swarms occur more commonly on summer nights, shortly after the first rain. Thousands of swarmers may emerge from numerous colonies at any one time, thus allowing intermixing of individuals from many populations. Termites are relatively weak fliers and their wings break easily. The presence of large numbers of wings from these swarmers is an indication that a colony is close by. The female attracts a nearby male and the two then search for a suitable nesting site. The first eggs are laid within a week to several weeks after mating and development of the colony is very slow for several years.

Figure 3-10. *Subterranean termite shelter tubes on the interior foundation wall of a crawl space type house.*

The usual first sign of a subterranean termite infestation is the appearance of the swarmers or discovery of the discarded wings, usually on a window sill or other lighted area. Another common sign is the presence of shelter tubes (**Fig. 3-10**) constructed over foundation walls or piers, in crevices between structural members, or on infested wood. When active infestations or shelter tubes are damaged, termite workers and soldiers will promptly appear (**Fig. 3-11**). Shelter tubes are used as a means for the termites to move from their nest in the soil to wood located above ground. These shelter tubes are made from particles of soil or wood and bits of debris held together with fecal material. The tubes do not conduct moisture, but rather serve to protect the termites from enemies and allow them to return to the soil.

The only other visible sign of an infestation is the presence of dark spots or blister-like areas on flooring, trim, or framing members. These areas are easily crushed with a knife or screwdriver. In cases of severe damage, a wood member might be partially collapsed at bearing points or otherwise failed. Internal damage can sometimes be determined by probing the wood with a sharp instrument or pounding with a hammer to detect hollow areas.

When termite-damaged wood is broken open, several characteristic features can be observed (**Fig. 3-12**). First, termites tend to eat the soft earlywood and leave behind the hard latewood. A thin outer shell is left around the entire wood member. Some cavities may contain substantial quantities of soil mixed with chewed wood. The gallery walls of the damaged wood and shelter tubes will have a pale, spotted appearance resembling dried oatmeal. The appearance is produced by the plastering of soft fecal material on the surface. There are no fecal pellets in subterranean termite galleries.

Several methods may be used to prevent and control subterranean termites. Treatment of the soil under and around the foundation with an approved insecticide is probably the most important step to take to protect buildings. Good design and construction practices should also be followed. Untreated wood should be at least 8 inches from the soil line and definitely not placed in contact with the soil. Earth-filled porches, flower planters, etc., should not be attached directly to the house. Wood debris should not be left or buried under or near the house. Pressure treatment of the wood with preservatives is another method of control. This simply makes the wood inedible for the insects. Therefore, termites are usually not a problem in preservative treated poles, piles, railroad ties, heavy timbers, and lumber unless they gain access through the protective preservative shell.

Where termites are a serious problem (**Fig. 3-13**), an inspection should be done yearly. Heavy damage from native subterranean termites is not likely to occur for the first 5 years after a house is constructed. When termites are found and damage is clearly not advanced, there is no reason for making hurried decisions. Substantial additional damage will normally not occur if proper treatment follows within 6 months.

Figure 3-11. *Subterranean termite workers.*

Figure 3-12. *Termite-damaged wood showing the insect's preference for the softer earlywood and accumulation of soil and/or fecal material in the galleries.*

Other Termite Species

Three additional termite species deserve mention, especially because of their significance in certain regions of the country.

Formosan Subterranean Termites. The Formosan subterranean termite has spread from the Far East to Hawaii, numerous other Pacific islands, California, and Texas to locations in the southeastern United States. This termite is expected to spread within its current range. Formosan termites multiply quickly and can cause serious damage in less time than the native species. If found, treatment within a few months is suggested. Although numerous differences exist among the Formosan termite and the damage they cause compared to the native species, the control measures given for the native subterranean species also apply.

Drywood Termites. Drywood termites occur in warm, humid climates ranging throughout the very southern United States from the East to West Coast (**Fig. 3-9**). In southern California, Arizona, and Florida, the Pacific area, and the Caribbean, new houses may be infected

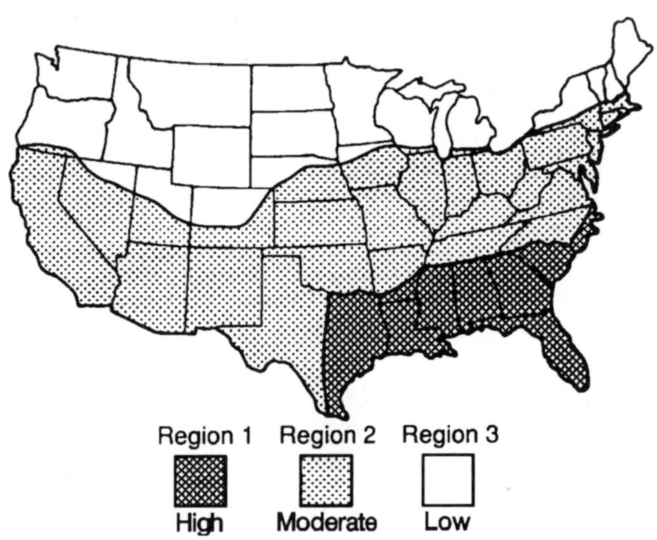

Figure 3-13. *Hazard regions for subterranean termites.*

within 5 years of construction. Drywood termites utilize the moisture within the wood they eat and do not need a source of water to live. These insects may colonize and damage smaller wooden articles such as furniture and packing materials; therefore, drywood termites can be important pests in wood items returning from the tropics.

Table 3-2. *Comparison of characteristics for termites and carpenter ants.*

	Subterranean termites				
Characteristics	Native	Formosan	Drywood termites	Dampwood termites[a]	Carpenter ants
Insect Description:					
Soldiers	Oblong, rectangular head	Oval-shaped head, exuded whitish, sticky substance		Very large, dark, rectangular head with large jaws	
Swarmers					
Color	Light tan to black	Yellow brown, larger than native subterranean	Slightly larger and lighter in color than subterranean	Yellowish brown to dark brown	Predominately black, some reddish-brown to yellowish
Size (length)	⅓ - ½ in.	Up to ⅝ in.	½ - ⅝ in.	Up to 1 in.	Up to ¾ in.
Waist	Not constricted	Not constricted	Not constricted	Not constricted	Very narrow, constricted
Wings	Four equal sized; two prominent longitudinal veins at front edge	Like subterranean but with hairy wings	Several distinct longitudinal veins with many cross veins	Veins predominately on front edge of wings, more numerous than subterranean	Two pair, unequal size
Other		Present in areas of TX, FL, LA, SC, and probably other southern coastal states	Will attack small wooden articles such as furniture		
Wood Description:					
Damage character-ization	Shelter tubes present; earlywood preferred; with pale, spotted appearance like dried oatmeal on gallery walls	Shelter tubes present; earlywood preferred; with pale, spotted appearance like dried oatmeal on gallery walls	Piles of fecal pellets eliminated from kickholes in wood; infested wood with very thin surface; broad pockets with no regard for earlywood or latewood; galleries smooth without deposits	Damage in sound wood is similar to subterranean damage. In decayed wood galleries are larger and pass through both earlywood and latewood; gallery walls with velvety appearance; some with dried fecal material	Galleries follow earlywood but will cut through late-wood gallery walls smooth and perfectly clean; "windows" or slit-like openings sometimes present on wood surface; sawdust-like frass and other debris present below windows
Moisture source or other conditions	Soil or other consistent supply of moisture required	Soil or other consistent supply of moisture required	No soil or moisture supply required	Damp or decayed wood required; soil or moisture supply not required	Damp or decayed wood preferred; will extend to dry, sound wood
Wood used for	Food	Food	Food	Food	Shelter

[a] Dampwood termites vary depending on species. This description is given for the Pacific Coast species, the only one of economic importance.

Drywood termites tend to work just under the surface of the wood, leaving a very thin veneer-like layer. Wood damaged by drywood termites has broad pockets or chambers connected by tunnels cutting across the grain without regard for earlywood or latewood (**Fig. 3-14**). Their galleries are perfectly smooth and have no surface deposits. Some fecal pellets may be stored in portions of the galleries, which are closed off by partitions made of fecal pellets stuck together with a secretion.

Piles of fecal pellets outside the wood are usually the first sign of a drywood termite infestation. The pellets are hard, elongated, less than 1⁄25 in. in length, rounded ends, and six flattened or concavely depressed sides, and are light gray to very dark brown in color. The pellets are eliminated from the galleries in the wood through round kick holes. When not being used, the holes are closed with secretion and pellets. Probing wood with a sharp instrument or pounding the surface may reveal hidden damage.

Methods available to prevent drywood termite attack are not as practical or as economical as those to prevent subterranean termites. First, any wood product or other cellulose-based material should be carefully examined before it is used or placed in a new or existing structure. Potential outdoor sources of infestation such as stored lumber, firewood, and dead branches should be examined or removed. The use of fine screening over attic and crawl space vents has been recommended, but this measure is not effective because the screen mesh must be so small to prevent insect passage that it would become clogged and restrict air movement. Desiccating dust application and using preservative-treated lumber are effective in prevention. The use of fumigants is effective for items in transit, but fumigating houses is expensive and may need to be repeated in areas of severe regional infestation.

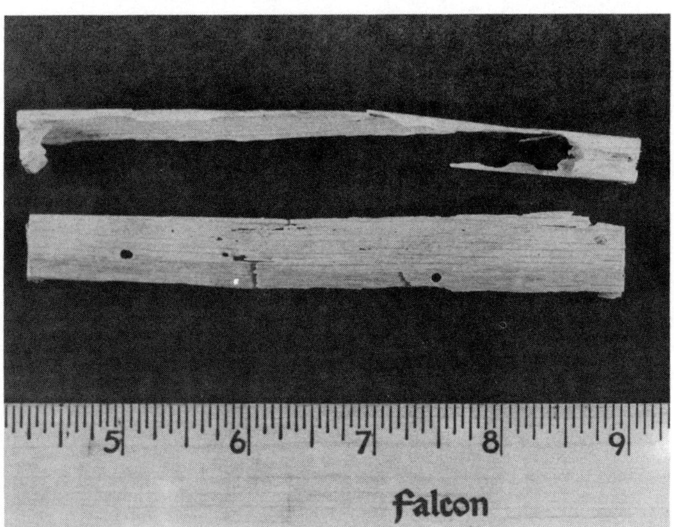

Figure 3-14. *Drywood termite damage showing round kick holes and the very thin veneer-like layer left after excavating without regard for earlywood or latewood.*

Dampwood Termites. Dampwood termites locate their colonies in damp and sometimes decaying wood and are of economic significance only on the Pacific Coast. Soil contact is not required if the wood is damp. Because these termites require the presence of damp wood, the same measures used to keep wood dry and prevent decay will control them. Dampwood termites are sometimes found in timber bundles shipped from the Pacific Coast.

Carpenter Ants

Carpenter ants are like termites because they live in colonies composed of several casts. There are winged and unwinged queens, winged males, and several sizes of unwinged workers. However, carpenter ants, like all ants, have a very narrow waist and wings of two different sizes, the front ones much larger than the hind ones. Adult termites do not have a restricted waist and have equal-sized wings. Ants and termites are compared in **Figure 3-15**. The adults found nesting in houses are predominantly black; however, some may be partially reddish-brown to yellowish.

Carpenter ants burrow into the wood to make nests and, unlike termites, do not use wood for food. Most species prefer to nest in moist hardwoods and softwoods

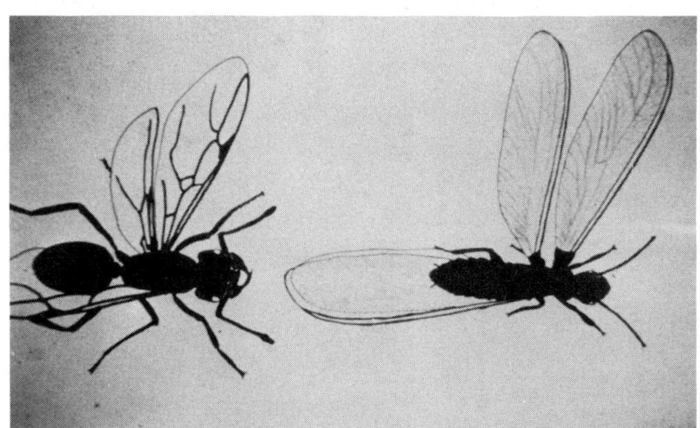

Figure 3-15. *Major differences between ants and termites include waist and wing size. The ant (left) has a restricted waist.*

that have begun to decay. The galleries extend along the grain of the wood and around the annual rings. The softer earlywood tends to be removed first. The harder-grained latewood is penetrated at frequent intervals, so that, unlike galleries made by subterranean termites, there is complete access (across annual rings) between the galleries (**Fig. 3-16**). When a nest is established, it can be extended into the sound wood. The surfaces of ant galleries are smooth and perfectly clean. This contrasts with galleries of subterranean termites where walls are rough due to the coating of fecal materials.

Carpenter ants can also nest in houses without attacking timbers. They simply use existing cavities, including wall voids, hollow flush panel doors, and termite galleries in wood. Carpenter ants foraging for food and water are generally more of a nuisance when found in the house.

In addition to the presence of large ants, other indications of infestation are piles or scattered bits of very fibrous and sawdust-like frass the ants have removed from the wood. If the pieces are from decayed wood, they tend to be dark and square ended. The frass will contain fragments of ants and other insects mixed with the wood fibers. The frass is expelled from cracks and crevices or slit-like openings called windows made by the ants. These windows are the only external evidence of attack by carpenter ants. The frass is quite often found in basements, dark closets, attics, under porches, and other out-of-the-way places. Faint rustling and even gnawing sounds can be heard in the wood or cavity when the ants are active.

Good building practices that keep wood dry will contribute to a reduced risk of attack by carpenter ants. These practices include adequate clearance between wood and soil, good drainage and ventilation, proper roof flashing, and tight exterior wood joints. If the wood cannot be kept dry, (e.g., porches, decks, columns), it should be pressure treated with an appropriate preservative. Chemical treatment of soil under and around a foundation to prevent subterranean termites does not prevent attack by carpenter ants.

If the good building practices just discussed have been followed and carpenter ants are present, the nests should be treated with residual contact insecticide applied as a dust or spray. The infested wood can be bored, the pesticide injected, and the holes plugged. The approaches and areas surrounding the nest should also be treated. Treating the areas where the ants are seen is seldom effective, because many ants never leave the nest. For treating indoors, pesticides specifically registered for that purpose must be selected. Poison baits may also be available.

Powder Post Beetles

The term powder post describes a group of beetles that convert the inner portion of infested wood to a powdery or pelleted mass. The thin outer shell on the surface of infested wood often shows numerous small holes where the beetles have exited. These exit holes aid in the recognition and identification of beetle attack. Damage is probably heaviest in warm, humid climates but can occur throughout the world. **Table 3-3** summarizes the major characteristics of these important wood-boring beetles as well as other less serious pests that may be mistaken for them. A detailed discussion of each group follows.

Anobiid Beetles. Anobiids, sometimes referred to as death watch beetles or furniture beetles, are found on recently seasoned and older hardwoods or softwoods throughout the United States. Sapwood, particularly that close to the bark, is preferred by anobiids. Unheated buildings or houses built with crawl spaces over damp ground, which often occur in the southeastern United States, are particularly susceptible, but houses are usually 10 or more years old before damage becomes obvious.

It is very difficult to detect an anobiid infestation. During and after emergence, a bright and light-colored powdery frass and tiny pellets will be found streaming from the exit holes or accumulating underneath infested wood. The frass in the galleries is loosely packed and does

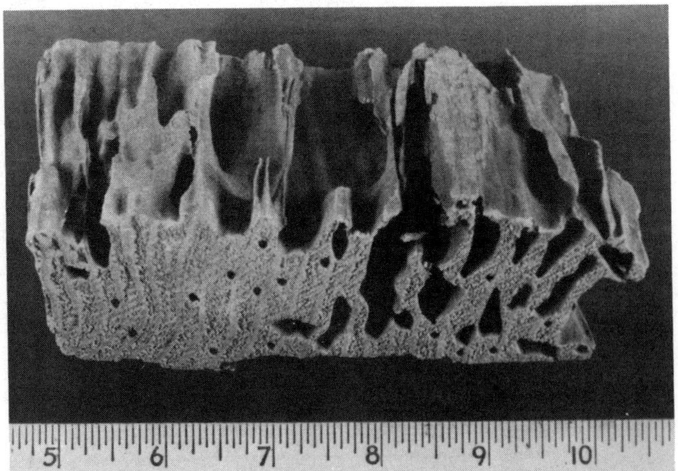

Figure 3-16. *Carpenter ant damage to Douglas-fir, showing preference for earlywood and removal of some latewood to allow access between galleries.*

not tend to fall freely from the wood unless the wood has dried out considerably since the attack occurred. The exit holes are round and vary in diameter from 1⁄16 to 1⁄8 in. (**Fig. 3-17**). When anobiid infestations die out naturally, the frass becomes yellowed and partially caked on the surface.

In most old, heavy infestations, very tiny round exit holes, about 1⁄32 in. in diameter, are scattered over the infested surface. These are emergence holes of parasitic wasps, whose larvae feed on the beetle larvae.

Lyctid Beetles. Lyctid beetles, also commonly called powder post beetles, attack the sapwood of hardwoods. They probably cause more damage to hardwoods than any other group of beetles and are a primary pest of imported hardwood products. Hardwoods with large pores such as oak, hickory, and ash are the most susceptible, but other species such as walnut, pecan, poplar, sweetgum, and cherry can also become infested. The greatest beetle activity occurs when wood moisture content is between 10 and 20 percent, but beetle activity can occur when the moisture content ranges from 8 to 32 percent.

Infested wood does not show any external evidence of attack until the first generation of adult beetles emerge. The circular emergence holes and longitudinal galleries are 1⁄32 to 1⁄16 in. in diameter (**Fig. 3-18**). Small piles of flour-like wood or frass can be found on or under the wood. The frass is loosely packed and a slight jarring of the wood will make the frass sift from the holes. With severe infestations, the sapwood may be completely converted to frass held in by a very thin veneer of surface wood with beetle exit holes. With heavy infestations, parasitic wasps, which make circular holes larger than those of the beetle, may be seen.

Control procedures for anobiid and lyctid beetles are based on three approaches: (1) eliminate or reduce the beetle population; (2) use a chemical treatment to protect susceptible wood or use naturally durable wood, if available; and (3) control environmental conditions in the wood to retard development of the beetles or larvae.

Proper building design will also help control beetles. Good ventilation in crawl spaces and attics and good drainage and proper clearance between wood and soil will help to reduce the equilibrium moisture content of wood in buildings; this creates conditions less favorable for beetle development. Before sawn wood is purchased or placed into service, it should be inspected. If an active infestation is suspected, the wood should be heat sterilized before leaving the manufacturing plant.

Pressure treatment with preservatives will prevent beetle attack. Diffusion treatments with boron will also protect wood from attack by beetles. Brushing or spraying wood with residual insecticides can also be used for control.

Fumigation quickly and completely controls infestations of wood-boring beetles whether the wood is in storage or service. In addition to acting rapidly, fumigants are useful when an infestation is extensive or in building locations that make other control procedures impractical. Because of the current uncertainty over the availability and use of residual insecticides and the large turnover in home ownership requiring certification that structures are insect-free, fumigation has become a common control procedure. Sulfuryl fluoride is the fumigant widely used

Figure 3-17. *Anobiid beetle damage showing insect exit holes. Zone lines indicating the wood is also partially decayed are evident on the end grain.*

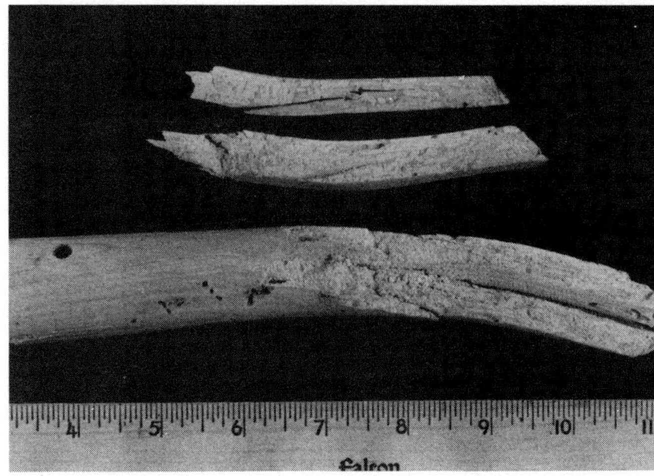

Figure 3-18. *Ash shovel handle damaged by lyctid beetles.*

for the control of wood-boring beetles in houses. Fumigation is an expensive process, and it offers no residual protection from reinfestation. In addition, it is necessary for the residents to leave the premises for 1 or more days.

Flatheaded Borers

A large family of beetles that attack trees and woody plants is commonly referred to as flatheaded borers because of the enlarged, flattened second and third thoracic segments of the larvae. A few species bore in seasoned wood. The winding tunnels made by the larvae are, in cross section, oval to flattened, about three times as wide as high, and packed with sawdust-like frass. Most infestations occur in trunks of weakened trees or in logs. The larvae survive in lumber cut from those logs. Kiln drying lumber is a good control procedure.

Other Wood-Boring Beetles

There are several additional groups of wood-boring beetles. Two are discussed here. These are the long-horned beetles, or round-headed wood borers, and the old-house borers. Collectively, both groups are called Cerambycids.

Long-Horned Beetles. Long-horned beetles are named because the mature beetle has antennae longer than half their body length. The larvae are described as round-headed wood borers, presumably because of their circular emergence holes (**Fig. 3-19**). Their damage may be mistaken as evidence of an active infestion of powder post beetles.

Damage from long-horned beetles usually occurs in fire-, disease-, or insect-killed timber that is being salvaged. The borers damage the wood before it is processed, and the activity ceases when the wood is dried. Many species feed just under the bark before moving into the sapwood and sometimes the heartwood. As beetles emerge, they make slightly oval to nearly round exit holds ranging from 1/8 to 3/8 in. or more in diameter. Most beetles emerge before the wood is 1-1/2 years old. If the damage occurred in the log, galleries in the sawn lumber may appear oval to elongate, depending on how the saw intersected the galleries. The texture of the frass varies from rather fine and meal-like in some species to very coarse and almost excelsior-like in other species (**Fig. 3-19**). Beetles are in the wood only a limited time and do not reinfest the wood; therefore, serious structural damage usually does not result.

Table 3-3. *Summary of exit/entry hole size, type, and condition of wood attacked, appearance of frass, and potential for reinfestation for wood-boring beetles and carpenter bees.*

Insect type	Shape and size of exit/entry hole (in.)	Wood type	Condition of wood attacked	Appearance of frass in tunnels	Reinfest
Anobiid beetles	Round 1/16 - 1/8	Softwoods, hardwoods	Seasoned	Fine powder and pellets, loosely packed (pellets may be absent and frass tightly packed in some hardwoods)	Yes
Carpenter bees	Round 1/2	Softwoods	Seasoned	None present	Yes
Flatheaded borers	Elliptic to flat, three times broader than high 1/4 - 3/8	Softwoods, hardwoods	Wounded or weakened trees	Packed with sawdust-like pellets	Generally no
Long-horned beetles (round-headed borers)	Round-oval 1/8 - 3/8	Softwoods, hardwoods	Unseasoned logs and lumber	Coarse to fibrous, mostly absent	No
Lyctid beetles	Round 1/32 - 1/16	Hardwoods	Newly seasoned	Fine, flour-like, loosely packed	Yes
Old-house borers	Oval 1/4 - 3/8	Softwoods	Seasoning to seasoned	Very fine powder and tiny pellets, tightly packed	Yes

Old-House Borers. The old-house borer is a round headed borer that, unlike most of the others in this family, is capable of reinfesting wood in use. It probably ranks next to the termites as the most important pest in the eastern United States. The heaviest infestations occur along the Atlantic seaboard, particularly the mid-Atlantic states. It is found in old and new construction, seasoned and relatively unseasoned softwood lumber, but not hardwoods.

The adult beetle is ⅝ to 1 in. long, slightly flattened, brownish in color with many gray hairs on its head and the forepart of the body. The borers are up to 1-¼ in. long. The adult lays its eggs in cracks and crevices in the wood. The larvae hatch in about 2 weeks. The larvae crawl over the surface of the wood and eventually bore their way into the sapwood. The larvae generally take 2 to 5 years to develop and, particularly in dry wood, larvae may take 12 to 15 years. A moisture content range of 15 to 25 percent encourages rapid development.

Early infestations are almost impossible to detect because there are no external signs. The larvae, when boring in the wood, make a rhythmic ticking or rasping sound much like that of a mouse gnawing. The borers tunnel very closely to the surface but seldom break through. The surface may bulge out because of the packed frass. After the adults have emerged, there may be small piles of frass beneath or on top of the infested wood. The frass, a very fine powder, is loosely packed in the tunnels. The exit holes are oval and ¼ to ⅜ in. in diameter. Serious damage can develop where enough humidity or moisture exists for reinfestation.

Figure 3-19. *Long-horned beetle or round-headed wood borer damage to ash, showing the tightly packed rather coarse frass in the galleries (top) and the circular exit hole (bottom).*

Carpenter Bees

Carpenter bees are found throughout the United States and are similar in appearance to bumblebees. They are more of a nuisance than a serious problem. The only external evidence of attack is the entry holes made by the females. The original entry hole is perfectly round and approximately ½ in. in diameter. The tunnel turns abruptly at a right angle after being extended approximately the length of the bee's body across the grain of the wood. The smooth-walled tunnel is extended with the grain from 4 to 6 in. in a new site (**Fig. 3-20**). When repeatedly used by succeeding generations, it may reach 6 to 10 ft in length. Frass accumulates on surfaces below the site of activity. Frass is usually the color of freshly sawed wood but varies in color with the species of wood under attack.

Carpenter bees usually choose wood that is soft and easy to bore such as redwood, baldcypress, cedar, white pine, and southern pines. Bare, unpainted wood is preferred; lightly painted or stained wood is also subject to attack. Carpenter bees will also tunnel wood that has been lightly pressure-treated for above ground use with waterborne preservatives.

Damage from carpenter bees can be prevented by keeping wood surfaces well painted. Interior, unpainted surfaces should be protected by keeping windows and doors closed or screened during the spring and early

Figure 3-20. *Carpenter bee damage in redwood showing a single entry hole (from inside) at the top center and individual cells (right).*

summer when the bees are looking for a nesting site. Wood pressure treated with organic preservatives, such as pentachlorophenol, or high retention levels of waterborne preservatives, is resistant to carpenter bee attack. Applying a surface coating of organic preservatives is also helpful.

Applying 5 to 10 percent carbaryl (Sevin) dust or any insecticide labeled for bee control into the entry holes will kill bees coming into contact with the residue. Several days after treatment, the holes should be plugged with short lengths of dowel rod or plastic wood. Plugging the holes without applying insecticide can lead to the production of new holes next to the plugs when bees inside attempt to emerge or nesting females seek re-entry into galleries in use.

Marine Borers

Three categories of marine borers damage wood in salt or brackish water. The first, and probably the most fascinating, are wood-boring shipworms. Most native shipworms belong to the genera *Teredo* or *Bankia* within the family Teredinidae. A second category is the Pholads or rock-boring piddocks of the family Pholadidae, most of which belong to the genus *Martesia*. Pholads, as they are commonly called, resemble clams. The last category is the crustaceans. These include the gribbles, genus *Limnoria*, in the family Limnoridae and the pillbugs, genus *Sphaeroma*, in the family Sphaeromatidae. The different types of marine borers are compared in **Table 3-4**.

Marine borers vary greatly in their distribution (**Fig. 3-21**) and ability to destroy wood. They are generally more destructive in tropical and subtropical waters. Their population depends on a number of factors such as salinity changes, water temperature, and dissolved oxygen content. There is some indication that marine borer populations increase as pollution levels decrease. Because of this variation, it is always best to consult local authorities for exact identification and before initiating control procedures.

Shipworms

Shipworms have a worm-like appearance and are notorious for causing extensive damage to wooden ships and other wooden waterfront structures. They are found in all coastal waters of the United States and some are capable of withstanding broad changes in salinity, thus they can also be found in the upper reaches of many estuaries. Boring is done by the mechanical rasping action of a pair of shell-hardened valves equipped with fine serrations on the head of the animal. The wood is partially digested to supplement the food furnished by other organisms. The posterior of the body has two siphons extending into the water. The incurrent siphon draws in water containing microscopic organisms for food and

Table 3-4. *Comparison of characteristics for marine borers.*

	Shipworms				
Characteristics	*Teredo*	*Bankia*	*Martesia* (Pholads)	*Limnoria* (Gribbles)	*Sphaeroma* (Pillbugs)
Appearance	Worm-like	See *Teredo*	Clam-like	Louse-like	See *Limnoria*
Area commonly attacked	Mud line	See *Teredo*	—	Between half and low tide, mud line; hourglass shape develops in piles	Between high and low tides
Damage	Entrance hole 1/16 in. or less in diameter; surface damage minimal; subsurface severely honeycombed holes lined with shelly calcareous deposits	See *Teredo*	Entrance hole 1/4 in. in diameter; damage close to surface; 2 to 2 1/2 in. deep, and 1 in. in diameter	Surface severely honeycombed, earlywood preferred; holes 1/16 in. in diameter by 1/2 in. deep; wave action erodes surface	Burrows typically 1/4 in. in diameter and about 3 times as deep, and perpendicular to the surface
Maximum borer size	Length 1 - 2 ft, diameter 1/2 in.	Length 5 - 6 ft, diameter 7/8 in.	Length 2 1/2 in., diameter 1 in.	Length 1/8 - 1/4 in., 1/3 as wide	Length 3/8 - 1/2 in., width 1/4 in.

dissolved oxygen for respiration. The excurrent siphon expels waste and reproductive products.

Entrance holes, about 1/16 in. or less in diameter, are bored into the surface of the wood by the young larva. Thus, it is difficult to detect their presence or the extent of damage from the surface of the wood. Attack is frequently concentrated at the mud line. The larvae enter the wood at right angles to the grain and burrow with the wood grain in the longitudinal direction, following a very irregular course, turning to avoid unfavorable sections such as knots. Inside the wood, the larvae rapidly develop worm-like bodies and remain imprisoned in the wood for life. The wood may become completely honeycombed (**Fig. 3-22**). Adults of the genus *Teredo* may be 1 to 2 ft in length and 1/2 in. in diameter, and those of the genus *Bankia* may be 5 to 6 ft in length and 7/8 in. in diameter. The wall of the hole is lined with a shell-like (calcareous) deposit. When the animal's siphons are withdrawn from the water into the hole, the opening is plugged with small calcareous pallets.

Shipworms are deterred by marine-grade creosote treatments of 20 to 25 lb/ft^3 of wood. Creosote should meet the current AWPA Standard P1/P13 or P2. Alternatively, 2.5 lb/ft^3 of wood of chromated copper arsenate (CCA) or ammoniacal copper zinc arsenate (ACZA) may be used.

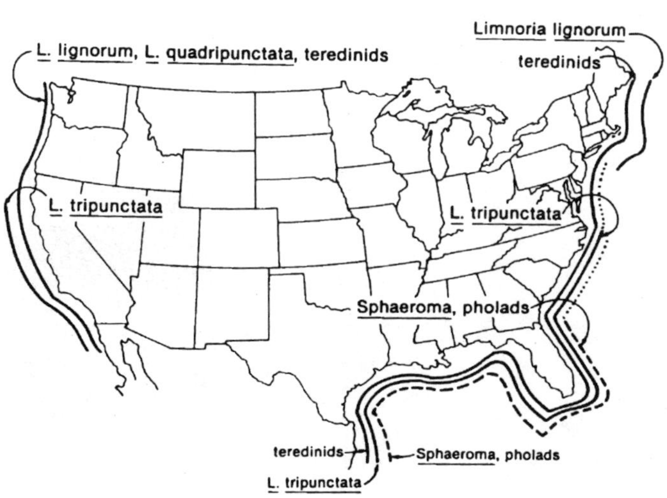

Figure 3-21. *Distribution of marine borers in the United States. Solid lines indicate areas where the designated marine borers are a hazard to wood that has not been treated with an appropriate preservative. The borers may be found beyond these ranges, but they have not been reported as causing problems. Dashed lines indicate areas where sporadic attack is probable. Dotted lines demarcate transition areas where problems with designated borers are unlikely but could occur.*

Pholads

Pholads resemble clams in that they have a pear-shaped body, which is entirely encased within the bivalve shell, even in adult form. These borers do not exceed 2-1/2 in. in length by 1 in. in diameter and become imprisoned within the wood. Pholads are found from South Carolina southward and become economically destructive in Florida and other tropical and subtropical waters. Pholads also cause serious damage in Hawaii and Mexico, and they are found to a limited extent around San Diego, California.

Pholads can bore in wood, clay, soft rock, shells, plastic, and even poor grades of concrete. The surface entrance hole is about 1/4 in. in diameter. The bore holes may be 2 to 2-1/2 in. deep and up to 1 in. in diameter. Thus, most of the damage is close to the wood surface. The diameter of the hole increases as the organism develops and increases in size.

Pholad attacks can be precluded by treating the wood with 20 to 25 lb/ft^3 of marine-grade creosote or creosote-coal tar or a dual treatment consisting of 1.0 to 1.5 lb/ft^3 of CCA or ACZA, followed by drying and treatment with 20 lb/ft^3 of creosote or creosote-coal tar. The CCA or ACZA alone are generally not effective.

Gribbles

Gribbles are marine pests that can cause extensive damage. However, their damage is generally visible and not as spectacular as the shipworms. One species, *Lim-*

Figure 3-22. *Damage from shipworms.*

noria tripunctata, found in temperate and subtropical regions, is tolerant of creosote in warmer waters. The other two species, *L. lignorum*, found worldwide in cool to temperate climates, and *L. quadripunctata*, found in temperate waters of the Pacific Coast, are not creosote tolerant. Gribbles are active along the entire coastal area of the United States. Their activity is accelerated in warm waters where breeding occurs year around.

Gribbles bore holes about 1/16 in. in diameter and seldom go more than 1/2 in. beneath the wood's surface. They may run obliquely for 1 in. or more. Gribbles prefer earlywood. Heavily infested wood may contain 300 to 400 gribbles/in^2, and the surface of the wood will be severely honeycombed. The remaining thin wood partitions are broken away by wave or other mechanical action and new wood is exposed to further attack. The attack is usually concentrated between half tide and low tide levels or at mud line where the eroding action is the greatest. A distinctive hourglass shape results (**Fig. 3-23**).

The method used to prevent *Limnoria* attack depends on which of the three species are present. This, in turn, depends somewhat on climatic conditions. *L. tripunctata*, which is destructive on the Pacific Coast from San Francisco southward, the Atlantic Coast from North Carolina southward, and the Gulf Coast, is tolerant of creosote in these regions. In temperate regions, it has been shown that a population explosion can occur if the water temperature increases slightly.

Two preservative treatments are recommended to control gribbles. One is a dual treatment consisting of treating pilings with 1.0 to 1.5 lb/ft^3 of CCA, drying the wood, and then treating with 20 lb/ft^3 of marine-grade creosote or creosote-coal tar. An alternative treatment is to use 2.5 lb/ft^3 of CCA. Treatments for *L. lignorum* and *L. quadripunctata*, found in temperate or cold waters, consists of using 20 to 25 lb/ft^3 of marine-grade creosote or creosote-coal tar or 2.5 lb/ft^3 of CCA or ACZA. See the current AWPA Standards for additional details.

Figure 3-23. *Extensive damage to a creosoted fender pile from marine borers, commonly known as gribbles (Limnoria).*

Pillbugs

Pillbugs, technically known as *Sphaeroma*, are similar to *Limnoria*, except pillbugs are longer and stouter, reaching a length of 3/8 to 1/2 in. long and 1/4 in. wide. Their burrows are typically 1/4 in. in diameter, generally perpendicular to the timber surface, and usually go to a depth of about three times the diameter (**Fig. 3-24**). Pillbugs prefer softer species of wood. Heavily attacked wood surfaces are often honeycombed. The wood is used for shelter and probably not ingested.

Figure 3-24. *Sphaeroma and associated damage.*

Wood-boring species of *Sphaeroma* are usually found in brackish waters along the south Atlantic and Gulf Coasts and, to a lesser extent, on the West Coast from San Francisco southward. They are most common in Florida estuaries, but their occurrence is spotty and unpredictable. They seem to be particularly tolerant of CCA and probably ACZA. Creosote seems to resist attack better, but with time becomes susceptible. Dual treatment with CCA or ACZA and creosote is the preferred method of protection against this borer.

Chapter 4
Protection of Wood

Protection of Wood

Degree of Protection Needed

The amount and type of protection needed for wood subject to attack by decay, insects, or marine borers will vary depending on several factors. Factors to consider are exposure of the wood and geographic location, permanency or expected service life, structural and nonstructural applications, and replacement costs. The types of protection available include using proper construction practices, naturally durable wood, and preservative-treated wood.

Geographic Location and Exposure

The four requirements of decay, food source, moisture, moderate temperature, and oxygen, were discussed in Chapter 3. The greater the decay hazard as shown in **Figure 3-3**, the more protection needed. **Figure 3-13** shows the hazard regions for subterranean termites. The most severe hazard is in the southeastern United States.

Another critical issue in determining the type of protection needed is whether the wood will be in soil contact or above ground. One of the most severe uses for wood is in contact with the soil. In this application, the conditions for decay and termite attack are the most favorable for the greatest length of time. Wood used above ground, as in fences and decks, does not have the usual constant source of moisture and it is usually exposed so that any insect attack is visible. As a result, the amount of preservative impregnated into the wood is sometimes decreased for lumber used above ground. However, because it is uncertain exactly how a piece of lumber will be used at the time of treatment, many companies will simply treat all lumber to a level adequate for soil contact. Such lumber is satisfactory for use above ground.

Marine applications represent a severe exposure hazard. Marine borers are some of the most difficult organisms to deter, and the potential for borer damage varies substantially with location. Thus, the quantity of preservative impregnated into the wood to prevent damage is increased for marine applications. In some cases, two types of preservatives are used. It is important to check with local authorities to determine the most appropriate preservative to use.

Permanency and Expected Service Life

In structures fully exposed to biological degradation and expected to provide a long service life, pressure-treated wood must be used. Examples include marine applications, wood railroad ties, and bridges and utility poles. In these commercial structures, the use of proper materials and treatments is so critical that companies who purchase them often retain their own or independent inspection agencies to ensure a quality product. Details are presented in Chapter 7.

Permanency is also a need in residential and agricultural applications, such as decks, sills, steps, pole barns, and posts. In these cases, the use of preservative-treated material is recommended as well. Steps in selecting a quality product are also presented in Chapter 7.

Structural and Nonstructural Applications

When wood is used in a structural application, particularly one where human life could be at risk, extra caution should be exercised. Some common examples in commercial applications include the piling, timbers, and ties used in a wooden railroad bridge or the large wooden poles used in a major electric transmission line. Another example using preservative-treated lumber might be something as basic as a cantilevered wooden deck on the second floor of a restaurant. In all these examples, the use of properly treated wood in a structure designed with the appropriate safety factors is critical if property damage and personal injury are to be avoided. Thus, when wood, especially preservative-treated wood, is used in structural applications, extra precautions should be taken. The importance of having the correct wood product properly treated with the appropriate preservative is so critical that specific standards have been developed for particular applications and special inspection procedures have been established.

Replacement Costs

Another issue to be considered in choosing the level of protection needed is the cost of replacing any material damaged from biological agents. In spite of current efforts, costs to replace damaged wood continues to run into the billions of dollars annually. Given labor costs, material costs, and the problems associated with the inconvenience of rehabilitation, it is difficult to imagine many long-term applications where the added cost of preservative treatment, where needed, is not justified.

Methods of Protection

Three general methods can be used to help protect wood from decay, insect, and marine borer damage: proper construction practices, naturally durable wood, and preservative treatment. The methods are not mutually exclusive. For example, it always makes good sense to use good construction practices with preservative-treated wood being used where needed.

Construction Practices

Untreated softwood lumber, plywood, and oriented strandboard are used in light-frame construction such as houses and agricultural and some commercial buildings. These structures should be designed to keep the wood dry and free from decay. Subterranean termites may be controlled by an insecticide applied to the soil. To keep wood dry in these structures, three basic premises must be strictly followed.

(1) All wood in or near the foundation, unless treated with a preservative, must be an adequate distance from the soil line to prevent it from becoming excessively wet.

(2) Starting from the roof line down, the structure must be built in such a way that any water falling on the roof or blowing against the side of the structure will run down and away.

(3) Promptly dispose of any moisture accumulated in the crawl space or water vapor generated in the living quarters. **Figure 4-1** shows a basic house with good construction practices that help to minimize moisture accumulation.

Wide roof overhangs provide some protection from sun and rain to at least the upper portions of the structure. When a 4-ft-wide overhang is provided, approximately two-thirds of a conventional one-story sidewall is protected from exposure to full sunlight. Any wood used for siding, sheathing, or plates should be at least 8 in. above the outside groundline unless pressure treated with a wood preservative.

Metal flashings under shingles at roof edges prevent water from entering the roof decking and sidewalls, particularly on roofs with a low slope that are located in areas of high rainfall. Metal flashings in roof valleys, junctions of roofs and walls, along dormers and siding material changes, and around chimneys and drip caps over window and door frames help prevent water from entering the house. Adequate and properly maintained gutters and properly hung downspouts prevent overflow and subsequent wetting of house eaves and sides and resulting rainwater splash of the siding near the ground level.

Adequate insulation and ventilation of attics and crawl spaces prevent moisture condensation and resulting high moisture content in the remainder of the house.

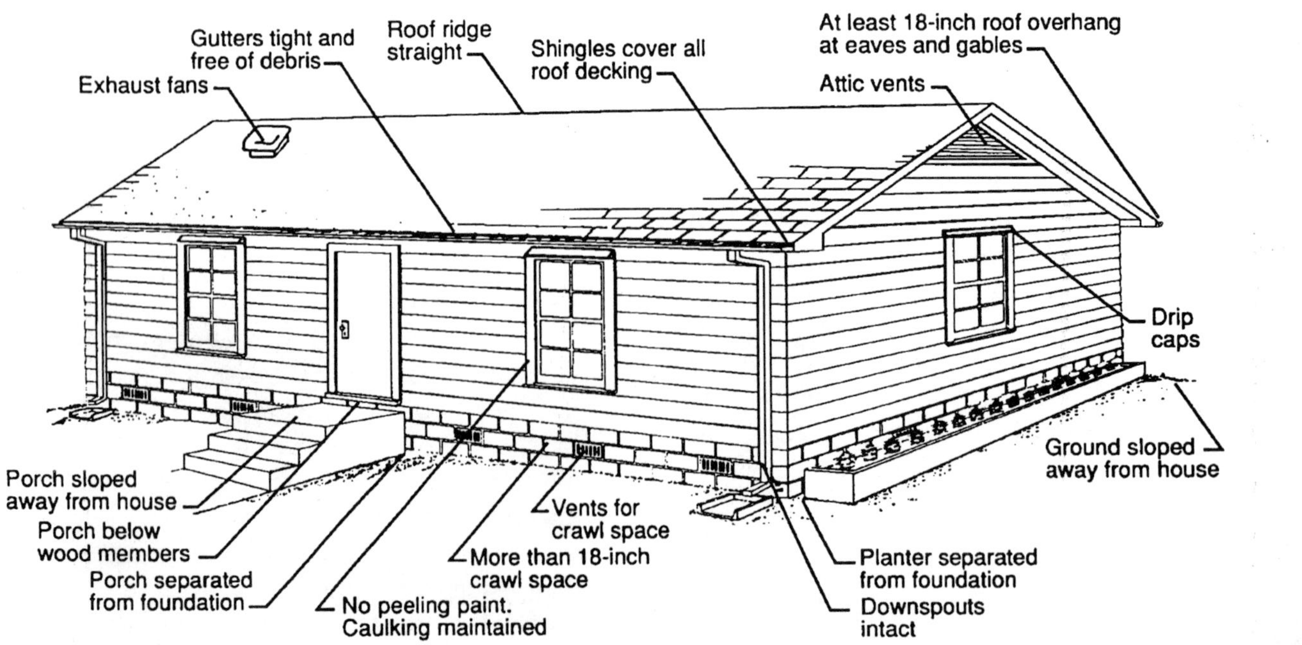

Figure 4-1. *Good construction practices prevent excessive buildup of moisture in wood members. Maintaining the correct moisture content in wood increases its service life and decreases the risk of wood decay and insect attack.*

Exhaust fans should be used to remove moisture from high humidity areas such as washrooms with showers or baths, kitchen areas, and appliances such as clothes dryers. Be sure the fans are vented to the outside of the house. Clothes dryers should never be vented to the inside living quarters or to the crawl space, basement, or attic. Plumbing should be well maintained.

If the house is built on a crawl space, a clearance of at least 18 in. between the soil and the floor joists is required, and wood girders must clear the soil by 12 in. Any standing water in the crawl space must be eliminated. The ground should be covered with a 4- to 6-mil thick polyethylene sheet or soil cover to prevent moisture movement from the soil upward. The crawl space should be adequately ventilated and the vents kept open, except in the coldest weather.

Proper installation of vapor barriers (vapor retarders), where appropriate, will help minimize condensation in floors, walls, and ceilings. Recent research shows that vapor retarders should be installed without air leaks around electrical receptacles and pipes. Traditionally, recommendations for vapor retarders (4- to 6-mil thick polyethylene sheets) have been based on climate. Three different climate zones are recognized in the United States, based on winter design temperatures (**Fig. 4-2**). Zone I includes areas with design temperatures <-20°F; zone II, between 0°F and -20 F; and zone III, >0°F, with the exception of areas with extremely warm and humid summers.

Vapor retarders are traditionally recommended for installation on the warm side of all exterior walls in zones I and II, and in zone III when the wall is insulated beyond R-4. Ceiling vapor retarders have traditionally been recommended for zone I only, but they are now recommended for zone II as well. These recommendations conform with current interim guidelines from the American Society of Heating, Refrigerating, and Air Conditioning Engineers.[3]

Humid climates deserve special attention (**Fig. 4-3**). Where dwellings are constantly air conditioned, warm moist air can move from the outside and condense on the cooler inside portion of outside walls. This situation is the reverse of cold weather condensation experienced in northern climates, but the same principle applies. In general, exterior surfaces of the building should be airtight and greater in vapor resistance than interior surfaces. That is, vapor retarders should be used on the outside of exterior walls. Any water that does enter the outside surface of the structure can flow to the inside where it can be removed by the air-conditioning system instead of accumulating in the floor, wall, or roof construction. The amount of condensation likely to occur is difficult to generalize and depends on local conditions.

Natural Durability

As indicated previously, the heartwood of some wood species is naturally durable or resistant to decay and insect attack. **Table 2-1** rates many common domestic species by their decay resistance, and **Table 2-2** rates several wood species commonly imported into the United States. Note that the only naturally durable native softwood species commonly available in construction lumber grades at retail lumber yards are the heartwood of cedars and redwood.

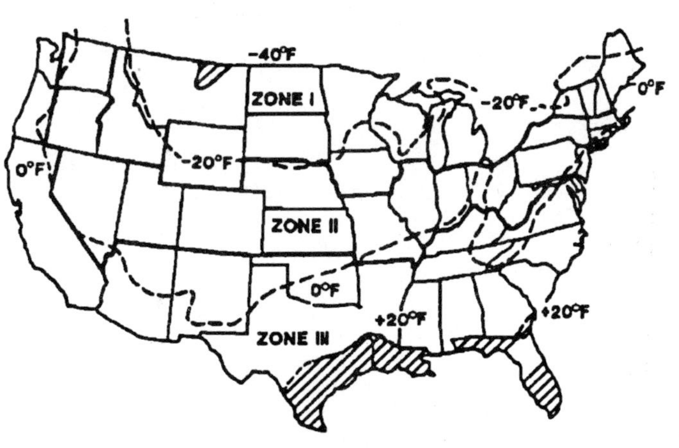

Figure 4-2. *Three different climate zones in the United States.*

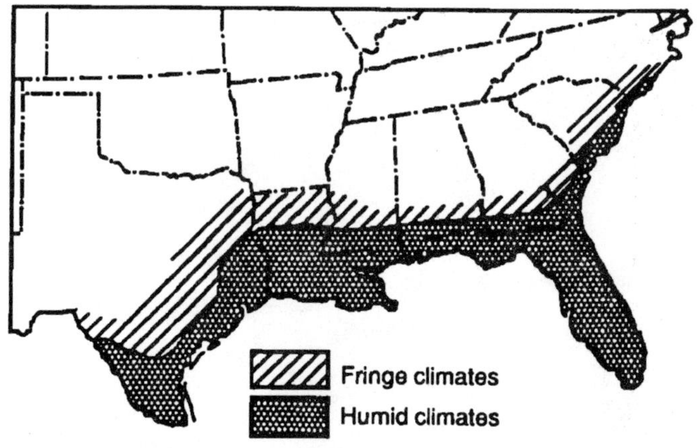

Figure 4-3. *Humid climates in the continental United States follow the coastal belt of the Gulf of Mexico and South Atlantic coastal areas. These areas deserve special attention concerning the placement of vapor retarders.*

Although the heartwood of redwood, cedars, and baldcypress has developed a reputation for natural durability, some caution needs to be exercised. This experience was developed using old-growth wood, and very little of this material is available today. Scientific studies using redwood[8] and baldcypress[5] have demonstrated that second-growth heartwood is not as durable as that of old growth. Both of these studies and a third study using western redcedar[23] have demonstrated that the natural decay resistance of these species increases with increasing heartwood diameter. Thus, small diameter second-growth timber generally has less durability. Of even more significance, these studies demonstrate that the natural decay resistance of timber is highly variable and cannot be controlled. Thus, it is best to apply the limited quantities of these species to decorative purposes or above ground applications where the decay hazard is less severe.

Of the naturally durable hardwood species, white oak is the most readily available in lumber form. Certain other species, such as black locust and osage-orange, are used for fence posts. When dealing with hardwoods, the low grades of lumber are sometimes sold for farm or other rough construction purposes. This material is generally affordable, but usually rough cut and not dried; as a result, it is not uniform. The heavier species are difficult to nail and saw compared with the standard softwoods. Despite this, at least some of this material continues to be consumed on a local use basis.

Wood Preservatives

To provide maximum protection, preservatives are used to protect wood. Numerous preservative treatment methods and standards exist and are the subject of discussion in the next chapter.

Chapter 5
Types of Preservatives

Types of Preservatives

Wood preservatives must be effective against wood-destroying organisms, but safe from an environmental and health standpoint for the end use intended. Numerous chemicals or chemical combinations have been used as preservatives in the past and many more have been proposed. The popularity of the different preservatives has changed depending on long-term experience and the preservative's effect on wood properties. In the past 15 years or so, emphasis has been on the effect of the preservative on the environment and any nontarget organisms that come in contact with the preservative. As a result, many newer products are entering the market and the availability of certain traditional preservatives is now limited. In this section, preservatives available to consumers at retail outlets are discussed. A discussion of preservatives most often applied in commercial operations follows.

Over-the-Counter Preservatives

Over-the-counter preservatives are those materials available to the general public. This is compared to restricted-use pesticides that require certification of the user. Over-the counter preservatives are most often applied by brush, spray, or dip treatment. Application methods and techniques are discussed in Chapter 6.

Water repellents applied by brushing, dipping, or spraying have been shown through research to be effective in reducing deterioration of wood from fungal activity in above-ground applications. Because they help wood repel water, they will also reduce checking and warping. Water repellents are mixtures of wax, resin or drying oil, and solvent. Water-repellent preservatives are water repellents with a preservative (fungicide) added. Both water repellents and water-repellent preservatives have been used for natural finishes and as wood stabilizers before painting. These materials are readily available to the general consumer and are often used to treat wood decks. Water repellents and water-repellent preservatives are designed to protect only the surface of the wood, and they must be renewed frequently, sometimes annually. They should not be confused with materials (preservatives) intended to provide long-term protection against biological agents. Preservatives available over the counter and those used in commercial pressure treating may also contain a water repellent to reduce wood checking and warping while the preservative controls biological activity. Be sure to carefully read the label or product information sheet to be certain that the material selected will perform as needed. The following presents some information on the more common over-the-counter formulations containing preservatives.

Copper naphthenate is a greenish-colored preservative that imparts the same color to the wood. However, the wood may turn brown after several months of exposure. Copper-naphthenate-based preservatives have typically been oil-based, but a water-based formulation is now available. For brushing and other superficial methods, a copper content of 1 to 2 percent (approximately 10 to 20 percent copper naphthenate) is usually required. This preservative is also used for pressure treating; additional information is provided in Chapter 6.

Zinc naphthenate is similar to copper naphthenate. It is less effective in preventing decay and mildew. It is lighter colored and does not impart the characteristic green color of copper naphthenate, although it does impart an odor. Waterborne and solventborne formulations are available.

3-Iodo-2-propynyl butyl carbamate, IPBC, (trade name Polyphase) is a wood preservative marketed by retail stores in several water-repellent-preservative formulations. It is intended for above-ground applications only. Some formulations can be painted, and no odor remains after drying. Waterborne and solventborne formulations are available.

Copper-8-quinolinolate solutions are green-brown, odorless, and toxic to a wide range of wood-destroying organisms, although it is not used where the degradation hazard is high. Because it has a low toxicity to humans and animals, it is permitted by the U.S. Food and Drug Administration for treatment of wood used in food handling. Examples include commercial refrigeration units, fruit and vegetable baskets and boxes, mushroom trays, greenhouse flats, and water tanks. It has also been used on nonwood materials such as webbing, cordage, cloth, leather, and plastics.

TBTO Tributyltin oxide or TBTO, technically known as bis (tri-N-butyltin) oxide, is a colorless to slightly yellow organotin compound soluble in many organic solvents, but insoluble in water. The TBTO is not effective when used in soil contact, and it is recommended only for above-ground use such as millwork. It has been used as a marine antifoulant, but its use has been virtually eliminated because of the environmental impact on shellfish.

Borate preservatives (borates), at least in the United States, are one of the newest preservatives to become available. They are derived from the mineral sodium borate; this mineral is the same material used in laundry additives. The preservative is effective against decay, termites, beetles, and carpenter ants when used in aboveground applications protected from wetting. Borate preservatives are odorless, do not corrode ferrous metal fasteners, and they can be finished with paint or other coatings. The preservative remains in place as long as it is not exposed to liquid water or does not come into prolonged contact with the soil. Borate solutions can be sprayed, brushed, or injected.

In green wood or wood that has become wet from exposure, borates will diffuse into the wood. Thus, in existing structures, this preservative will diffuse into those areas that are wet and most susceptible to decay. Because the preservative diffuses into the wood, it also has the potential for treating large timbers or other wood members that are not easily pressure treated. Borates are widely used in the log home market and several natural wood finishes. The material has also been used to treat hardwood pallets where a bright appearance and absence of insects is important. Remember that this preservative is leachable.

Borates have been made available to the pest control industry and sometimes directly to the consumer. Because borates have been recognized in AWPA standards for Southern Pine lumber not in contact with ground or liquid water, the number of commercial users and companies marketing the product is expected to increase.

Commercially Applied Preservatives

Many currently used preservatives for pressure treatment of wood have been available for several years. As a result, actual-use data on long-term durability are available. **Table 5-1** presents fence post test data on four of the six preservatives presented in this section. Some of these preservatives have extended the service life of wood posts by at least tenfold when compared with untreated posts.

Preservatives applied by commercial pressure-treating operations can be placed in two classifications: oil-type and waterborne preservatives. The traditional oil-type preservatives are creosote, pentachlorophenol (often called penta), and recently, copper naphthenate. The traditional waterborne preservatives are chromated copper arsenate (CCA), ammoniacal copper zinc arsenate (ACZA), and recently, ammoniacal copper quaternary ammonium compound or ammoniacal copper quat (ACQ). With the exception of copper naphthenate and ACQ, the preservatives are restricted-use pesticides. That is, they can only be purchased and applied by individuals who have received specific training on the safe handling and use of pesticides. However, the treated wood is not restricted and is available for general consumer use. Information on the safe use, handling, and disposal of the wood is given in Chapter 11.

Oil-Type Preservatives

Creosote, or coal-tar creosote, is the one preservative that has persisted since the earliest days of preservative treatments. In the United States, it was first used for marine pilings in 1889 and, by the 1920s, it became the treatment of choice for the railroad industry and continues to be so. Creosote is also commonly used on heavy timbers, poles, and piles.

Table 5-1. *Examples of expected service life for Southern Pine posts in soil contact and pressure treated with different preservatives.*[10] *The posts were installed in southern Mississippi in an area with a high decay hazard rating and abundant termite populations. Because treatment and retention varied considerably, these data do not represent a comparison of preservatives. Rather, they are intended to show the large increase in service life that can be achieved with different preservatives.*

Preservative	Retention (lb/ft^3)	Estimated service life (year)
Chromated copper arsenate[a] (CCA) - Type C	0.41	26+[b,c]
Coal-tar creosote[a]	5.6	46
Copper-8-quinolinolate (1.0% in petroleum oil)	8.1	26+[b,c]
Copper naphthenate (0.5% in petroleum oil)		50[b]
Pentachlorophenol[a] (5% in petroleum oil)	6.3	50
No preservative treatment	-	3.3

[a] Restricted-use pesticide.
[b] Test still in progress.
[c] No failures as of 1990.

Creosote is distilled from coal-tar. Coal-tar is a by-product derived when coke is produced by the high temperature carbonization of bituminous coal in sealed airtight ovens. Creosote is effective against most biological organisms that attack wood. It imparts a dark brown to black color and has a characteristic odor. It is permanent due to its low volatility and solubility in water. The odor and oily dark surfaces of the treated wood may be objectionable in some applications and freshly creosoted wood may ignite easily. Creosoted material cannot be painted and should not be used indoors or where human contact is likely. In November 1986, creosote became a restricted-use pesticide and it is available only to certified applicators. However, it is available as a General Use Preservative for contractors. For use and handling precautions for creosoted treated wood, refer to the Consumer Information Sheet provided by the U.S. Environmental Protection Agency (EPA) in Chapter 11.

Pentachlorophenol has been widely used for both pressure and nonpressure preservative treatment of wood. In heavy oils, common applications include poles, posts, timbers, and lumber. In this form, the material leaves an oily residue on the surface that cannot be painted. Odor, except for that of the oil carrier, is not a problem. When light oils are used as a solvent for pentachlorophenol, the treated wood is typically much lighter in color than creosoted material.

Pentachlorophenol is effective against many terrestrial wood-destroying organisms such as mold, stain, decay fungi, and insects. However, it is not used in salt water because of its leachability and limited efficacy. It should not be used indoors or where human, plant, or animal contact is likely.

Pentachlorophenol in a 5-percent solution of light petroleum solvents, such as mineral spirits, has been used for treating exterior millwork, because the solvent evaporates and leaves a clean, paintable surface.

Pentachlorophenol was one of the most common preservatives for commercial pressure treatment as well as residential and agricultural use until oil prices escalated in the early 1970s. It became a restricted-use pesticide in November 1986. During the same period, use of the waterborne preservatives that leave an oil-free and paintable surface increased markedly. As a result, pentachlorophenol has lost its popularity for certain products. It is found in many older installations and still used in substantial quantities for treatment of utility poles, crossarms, and bridge timbers. It is commonly used as a standard to compare the effectiveness of other or newer preservatives. For use and handling precautions, refer to the Consumer Information Sheet provided by EPA in Chapter 11.

The toxicity of copper naphthenate to wood-destroying fungi and insects has been known for close to a century, and the preservative has had commercial use since the 1940s. It is a mixture of naphthenic acids, usually obtained as by-products in petroleum refining, and copper salts. Use of this preservative for poles and other applications to protect against decay and insects has increased as a replacement for pentachlorophenol.

Copper naphthenate is a dark-green liquid. The treated wood color may vary from olive to dark green when treatment is done at ambient temperatures. Treated wood generally weathers to a light brown after several months of exposure. If heat is used in the treating process, the wood may vary from light brown to chocolate. The material is noncorrosive and leach resistant when applied to wood.[13]

Copper naphthenate is not a restricted-use pesticide. However, it should be handled as industrial pesticide. Handling precautions are similar to those for wood treated with oil-based and waterborne preservatives. Be sure to obtain current detailed information from the supplier.

Waterborne Preservatives

Lumber and many other types of wood products including posts, poles, and some marine pilings are commonly treated with one of several waterborne preservatives. The resulting wood product is usually somewhat green or brown, and no odor results after seasoning. The surface of properly treated wood is free of excess preservative, and the dry wood can be painted.

The more common waterborne preservatives include chromated copper arsenate (CCA), ammoniacal copper zinc arsenate (ACZA), and ammoniacal copper quaternary compound (ACQ). There are two different forms of CCA currently used, both of which have performed well against decay and insect attack. Type C is by far the most common form of CCA currently being used. In addition, high retention of 2.5 lb/ft^3 of CCA will provide good resistance to the marine borers *Limnoria* and *Teredo*.

Ammoniacal copper zinc arsenate (ACZA) is used almost exclusively on the West Coast for the treatment of Douglas-fir. The chemical makeup and stability at elevated treating temperatures of this preservative improve its penetration of refractory Douglas-fir heartwood.

Wood treated with ACZA is similar in performance and most characteristics to that treated with CCA.

New waterborne wood preservatives that do not contain either arsenic or chromium are being developed. A variety of formulations are being proposed and treatments that are suitable for a number of different woods including southern pines and Douglas-fir are anticipated. As these new preservatives pass the necessary review steps to ensure reliable performance, their appearance in the market place will be under new trade names and acronyms, but the quality control process will be the same as that which has been employed with CCA and ACZA treatments. Ammoniacal copper quaternary compound (ACQ) Type B, is an example of a new waterborne preservative. Copper citrate and copper dimethyldithiocarbamate are among other waterborne formulations that will likely be recognized in future AWPA standards.

Properties of Treated Wood

Shrinkage can be an important factor when using wood treated with the waterborne preservatives. When wood is treated, it is injected with large quantities of water that carries the preservative chemical. The water swells the wood to near its original green dimensions. Thereafter, lumber may seem dry on the surface, but be quite wet in the interior. If the wood is installed while it is still wet and allowed to air dry as in an exposed deck, it will shrink about 3 to 5 percent. This dimensional change can amount to about 1/4 in. on a nominal 6 in. wide board. Thus, if the spacing between boards or the dry dimension of the structure is critical, the amount of shrinkage should be considered when using undried wood (**Fig. 5-1**).

Figure 5-1. *Tongue-and-groove lumber treated with a waterborne preservative and not dried before installation was used for this deck. The tongue-and-groove boards were laid tight against each other. Note the shrinking and warping that can occur.*

Warping and twisting as wet wood dries are also common. Thus, the wood should be firmly attached so it will remain flat or it should be airdried under restraint. The use of spiral or ring-shanked nails will help hold the pieces in place.

Corrosion of metals used in contact with wood treated with the oil-type preservatives is usually not a problem. Heavy oils tend to coat the surface of the fasteners. In wharf and shoreline bulkhead construction where metals are exposed to salt water, hot-dip galvanized steel is usually recommended. Corrosion of metal fasteners within the wood is inhibited in these applications because the holes are most often drilled prior to pressure impregnation of the preservative.

Corrosion of metal fasteners can occur as a result of the presence of small quantities of soluble copper ions in the waterborne preservatives. Under conditions of low moisture content, corrosion will be negligible. When the moisture content of the treated wood approaches 20 percent, corrosion can occur. In above-ground applications, hot-dip galvanized steel is acceptable to use for fasteners. When moisture is commonly present, as in soil contact, stainless steel is most often recommended as the appropriate fastener. In applications where structural integrity or life-endangering situations could develop, stainless steel is also preferred.

The strength characteristics of wood treated with waterborne preservatives should also be considered. Waterborne preservative treatment of wood produces a clean, odor-free, paintable or stainable product. However, waterborne preservatives can reduce wood strength. The extent of this effect depends on several key factors such as preservative chemical, redrying temperature, species, grade, size, and incising. Studies have shown that ACA and ACZA treatments affect the strength of wood less than CCA treatments. All terrestrial retention levels appear to have similar effects on wood strength when redried at comparable temperatures. Redrying temperature appears to be the most decisive, single processing factor affecting strength. It appears that no differential species effect exists. Incising reduces strength. However, this strength loss is more than offset by the improved performance of the incised treated product. Existing standards have been modified to reflect these new findings.

Chapter 6
Preservative-Treatment Processes

Preservative-Treatment Processes

Preservative-treatment processes can be divided into two broad categories. The first category consists of superficial nonpressure treatments including brushing, spraying, dipping, or soaking. With few exceptions, these methods are most often applied on-site and offer minimal protection. Vacuum, diffusion, and hot and cold bath treatments can improve penetration somewhat over the strictly nonpressure methods. Pressure-treating processes involve different combinations of vacuum and pressure and are done in commercial installations. Pressure treatments are capable of offering the maximum protection available.

Requirements for an Effective Wood Preservative

The degree of protection offered by a particular preservative and treatment process is dependent upon the fulfillment of four basic requirements. These are the toxicity and permanence of the preservative to potential biological agents, the amount of preservative impregnated into the wood (also called retention), and the degree of penetration of the preservative into the wood.

Toxicity refers to how effectively a preservative deters decay fungi, wood-destroying insects, and marine borers. Some preservatives, or particular formulations or mixtures of preservatives, are more effective than others. The AWPA standards define applicable preservatives.

Permanence is also required for an effective preservative. The preservative must be resistant to leaching and volatilization, as well as chemical and biological breakdown.

Retention is the term used in wood preservation that specifies the amount of preservative that must be impregnated into a specified volume of wood to meet standards and ensure that the product will repel the numerous biological agents. Some preservatives are more toxic than others, and some biological organisms are more tolerant than others. Therefore, the recommended retention can vary depending on the preservative and end-use intended. Retention levels, usually expressed as lb/ft^3, are set by AWPA standards and should be strictly followed.

Penetration of the preservative into the wood is the fourth major requirement for effectiveness. As wood is pressure treated, the preservative penetrates from the ends and sides of the piece. The end grain is relatively easily penetrated, and the sides are more difficult to penetrate. Some wood, such as the southern pines, are easily treated because the wood is predominately sapwood, and others like Douglas-fir, which is predominately heartwood, are very difficult to penetrate. Using even the best pressure-treating technology, it is difficult for a preservative to penetrate most woods more than a few inches. As a result, the interior is not treated in timbers, poles, and other large members. The outer-treated wood forms a protective shell around the untreated interior. This protective shell must not be broken or the interior of the member could be attacked. Penetration levels, generally expressed in inches, are set by AWPA standards and should be strictly followed. When these standards call for incising of refractory or difficult to treat species, incising should not be waived.

Nonpressure Treatments

Brushing, spraying, dipping, or soaking are all nonpressure processes that, with the exception of soaking in certain situations, only provide superficial protection to wood. The preservative does not penetrate deeply into the wood, and the treatments are not intended to protect the wood from sustained exposure to degrading organisms. These treatments should only be used for above-ground applications where the decay hazard is relatively small or field treatment of exposed surfaces where pressure-treated wood is cut. Some preservatives and their solutions can pose a health hazard to the applicator as well as resulting in environmental damage. Therefore, read and follow all label directions and exercise caution when working with these materials.

Brush treatment is probably the simplest of all treatment methods. The preservative solution is simply flooded onto the surface and brushed out. Brush treatment is preferred to roller or spray application because the movement of the brush helps introduce preservative into the micro-fissures of the wood. Checks and other openings should be saturated to the point of refusal. A second or third treatment after the first has been absorbed will improve the effectiveness. Wood should be dried well before treatment or it will not accept preservatives applied in this manner.

When preservatives are sprayed, they can become airborne and present a potential hazard to the applicator and the environment. Within factories, these concerns can be addressed through engineering design and maintenance of operating equipment. When preservatives are to be applied outdoors by this method, a coarse spray should be used and the operator should wear protective clothing and respiratory equipment. Special precautions should be taken to avoid spray drift.

Dip treatment involves the immersion of wood materials into a tank of preservative until the wood absorbs the appropriate amount of solution or until it refuses to accept additional preservative. This method is used commercially to treat millwork with a water-repellent preservative. A 3-minute submersion of permeable wood species has been shown to give substantial protection against moisture absorption and subsequent decay in millwork used above ground (**Fig. 6-1**).

The National Wood Window and Door Association (NWWDA) maintains a standard for the nonpressure treatment of wood. The standard requires that any water-repellent preservative formulation be registered by the Environmental Protection Agency and that the formulation effectively retard molds, fungus stains, and decay. The formulation should not discolor the wood, cause objectionable surface residue, or have an objectionable odor, and it should be compatible with other materials used in the fabrication of the treated product. Normal shrinking and swelling of wood should be reduced by at least 60 percent when exterior millwork is treated with a water-repellent preservative. The AWPA is in the process of establishing standards for nonpressure treatments for wood used above ground.

Cold soaking is simply the soaking of bark-free dried wood in an unheated oil-based preservative such as creosote or pentachlorophenol. Cold soaking well-dried, permeable woods, such as the sapwood of pine, for 2 days to 1 week may result in absorption of 4 to 6 lb/ft^3 of preservative. Thus, with little equipment cost and effort, reasonable protection against decay and insects can be achieved, especially for small, round products such as fence posts. Care should be taken to avoid soil contamination by spills or drips from freshly treated products.

Other Treatments

For vacuum treatment, the air in a chamber is exhausted with a vacuum pump and the preservative is introduced. After the preservative enters the chamber and the vacuum is released, air at atmospheric pressure increases penetration.

The hot and cold bath is also known as the open-tank treatment, the boiling-and-cooling method, or the thermal process. Creosote and other oils are commonly used with this process because large quantities are not lost through heating and evaporation in open tanks. However, water-soluble salts have also been used effectively. The process is used for butt, as well as for full length, treatment of airdried, bark-free poles. Butt treatment protects the most vulnerable part of the pole. The process is capable of treating the nondurable sapwood portion. As a result, it has been used in those sections of the country with a low decay hazard rating and where cedar poles with naturally durable heartwood are used. It has also been used on fence posts, lumber, and other timber products, particularly when small quantities are treated.

Figure 6-1. *Contrast between dip-treated (top) and untreated (bottom) steps after 5 years of exposure near Gulfport, Mississippi. On-site dip treatment consisted of immersing the precut lumber 3 minutes in a light oil solution of 5 percent pentachlorophenol and water repellent. Pentachlorophenol is now a restricted-use pesticide but other preservatives are available.*

Material that is to be treated by the hot and cold bath method must be thoroughly dried. Dry wood is first immersed in a hot bath of preservative for about 6 hours. The temperature of creosote or pentachlorophenol for the hot bath should not exceed 230°F. The hot bath expands the air in the cell lumens and forces a portion of it out of the wood. The hot bath is followed by a cold or cooling bath. In the case of the cold bath, the hot preservative is replaced with a preservative at a lower temperature. With a cooling bath treatment, the hot preservative is left in the treating tank and allowed to cool. As the wood cools, air and vapor in the outer cell lumens contract, forming a partial vacuum, and atmospheric pressure forces the preservative into the wood. The cold bath temperature is usually between the point where solids form in the preservative and 150°F. It normally lasts 2 hours or more. A final expansion bath of hot preservative oil may be applied at the end of the treating process.

Diffusion is a method of treating green wood where the chemical preservative moves from an area of high concentration (the chemical at the surface) to an area of lower concentration (inside the wood) over extended periods. This method has been used for years to a limited extent in wood treating. One successful and continuing application is the remedial treatment of poles at groundline. In this case, the soil is removed from the groundline area, preservative is applied to the pole (after all decayed wood has been cut away), and the pole is then covered with a protective wrapping. The hole is back filled. Active ingredients slowly diffuse into the high moisture content wood to suppress growth of decay fungi and resist insect attack.

With the introduction of borates as preservatives into the U.S. market, the use of the diffusion process may expand into other areas. Borates readily diffuse into wet wood. However, when exposed to liquid water, they will diffuse back out. That is, they are not permanently locked into the wood. Applications will be effective in locations where wood has become wet or subject to insect attack, but where it is not exposed to the rain or other free moisture sources. Examples might include wood in crawl spaces or cabin logs. Recent research has shown that borates are effective preservatives, not toxic to humans, and can be effectively applied by diffusion on green or wet wood.

Double diffusion is similar to diffusion except two different chemical solutions are used. In this process, the green wood is first soaked in one chemical solution such as sodium fluoride and then in a second solution such as copper sulfate. The chemicals combine to form a leach-resistant compound, copper fluoride. Copper sulfate and sodium chromate with sodium arsenate may also be used. The double diffusion method can be an effective method of wood preservation and, unlike pressure treating, requires little capital equipment. However, the treating solutions are very toxic and can create serious safety problems in handling and disposal. Consequently, this process is seldom used. It is very important to follow EPA guidelines for all stages of use and disposal of the treating chemicals.

Pressure Processes

In commercial pressure treating, three general areas of processing must be addressed. These are the preparation of the material before treatment, the preservative treatment itself, and any conditioning that must be done after the treatment process is finished. Each of these three areas can vary substantially depending on the type of product (e.g., lumber, poles, piles, timbers) being processed, the intended end-use, and the preservative being used. A general description of the processes follows.

Pretreatment

Bark on wood prevents it from drying, blocks the penetration of preservative into the wood, and encourages wood boring and decay organisms to develop. Therefore, all bark must be removed, especially from round products. For poles and posts, shaving machines are normally used. In the spring of the year when growth first begins, bark is more easily peeled and hand equipment is sometimes used on small quantities.

Next, the wood must be dried before treatment to obtain satisfactory penetration and retention of preservatives. Air drying is the simplest method. Wood is stacked with adequate room for air to circulate around each piece, thus allowing water to evaporate. Wood products are also dried in kilns. Dry kilns are simply large ovens used to accelerate the drying of wood without producing excessive defects. Temperature, humidity, and air velocity in the kilns are controlled.

Another method for drying wood to be pressure treated is steam conditioning. Wood in a treating cylinder is subjected to live steam at 220°F to 245°F for 1 to 20 hours followed by a vacuum. Moisture is removed during the vacuum period. Unfortunately, steaming can also reduce wood strength. Therefore, the maximum allowable steaming times and temperatures are defined in AWPA standards.

Boultonizing is a drying method also applied in a treating cylinder by heating wood in creosote or an oil-type preservative while subjecting it to a vacuum. Water will boil at less than 212°F because the wood is under vacuum. Therefore, this process uses temperatures less than those required in the steaming and vacuum process.

After air or kiln drying, certain species and products may be incised. Incising is the production of small slits or holes in the surface of wood products to facilitate deep and uniform penetration of preservative by creating additional end grain surface area (**Fig. 6-2**). It is commonly used on woods such as Douglas-fir, Hem-Fir, and spruce that do not readily accept preservatives. Incising requirements for different species and commodities are given in AWPA standards.

Finally, all machining and cutting, such as boring or framing of pole tops, should be done prior to preservative treatment to reduce exposure of untreated portions of wood to decay fungi, insects, and marine borers.

Treating

Pressure-treating processes are the most commonly employed method to obtain deep penetration of wood preservatives. These treatments are done in retorts, or closed cylinders, where penetration and retention can be controlled. These cylinders are usually 4 to 10 feet in diameter and up to 180 feet long (**Fig. 6-3**). Preservative storage and measuring tanks, pressure and vacuum pumps, steam boilers, measuring and recording instruments, and other equipment are usually located adjacent to the cylinder.

Pressure processes can be divided into two categories: full-cell and empty cell processes. A general description of each process follows.

The full-cell process is also called the Bethell process and was patented in 1838. This process is designed to force as much preservative into the wood as possible. The cell lumens, or voids, are loaded with preservative, resulting in a high level of retention. This process is commonly used with creosote treatments for marine piling where high retention levels are required and with the waterborne preservatives. With oilborne treatments or creosote, the full-cell process forces so much preservative into the wood that some of it may bleed onto the surface after the treated-wood products are installed. Therefore, with these preservatives, this process should only be used when additional retention is needed. The modified full-cell process is a relatively recent development intended to accelerate post-treatment conditioning of wood products that have been treated with waterborne preservatives.

Figure 6-2. *Incising improves the treatability of refractory woods such as Douglas-fir.*

Figure 6-3. *Typical treating cylinder with charge ready for processing.*

The empty cell process impregnates and coats the cell walls with preservative. The cell lumens are not filled. This is accomplished by using a slight initial pressure within the wood product to force excess preservative out of the wood at the end of the treating cycle. This process provides the deepest penetration with low retention of preservative. The Reuping and Lowry processes are the two empty cell processes most commonly employed. Exact specifications for products produced by both full- and empty cell processes are given in the AWPA standards.

Post-Treatment

After the wood has been pressure treated with a preservative, two additional steps can be taken depending on end-use requirements. These are cleaning or steaming the surface to remove excess preservative and fixing the waterborne preservative to increase its leach resistance. Both cleaning and fixing may be done in the treating cylinder. To decrease processing time in the treating cylinder, kilns may be used to dry the wood to a specified moisture content.

The cleaning process is important. With oil-based preservatives, it removes excess preservative, thus improving appearance and facilitating handling. With the waterborne preservatives commonly used where human exposure is likely, surface cleaning will reduce the amount of chemical on the surface (see Chapter 11, Safety and Environmental Issues).

The redrying of wood products treated with waterborne preservatives is important for two reasons. First, dry wood is lighter and stronger and it will not shrink, warp, or check as severely as will wood installed when it is wet. However, most standard lumber treated with the waterborne preservatives is not kiln dried before shipment, and the material should be installed understanding that shrinkage is likely. Secondly, and more importantly, the fixation process for chromium-containing preservatives is completed when wood is heated and dries either in the atmosphere or dry kilns. During fixation, hexavalent chromium converts to its trivalent form and a leach resistant (fixed) complex is formed. The fixation period is extended by cool temperatures. The fixation process can also be completed before the wood is removed from the treating cylinder. The current AWPA Standard C1 permits hot water or steam to be injected into the cylinder to accelerate the fixation process as long as the temperature does not exceed those set in the treating standards.

Chapter 7
Standards, Quality Assurance, and Specifications

Standards, Quality Assurance, and Specifications

The extent to which the quality of preservative-treated wood products needs to be monitored is directly related to the project being constructed. Minimal concern will probably exist when a dozen landscape timbers are being purchased to construct a raised rose bed or where a few boards are being purchased for a wood deck constructed a few inches above ground level. On the other hand, maximum concern will be expressed and all precautions will be taken if a utility company is constructing a high voltage cross country transmission line using large wood poles or a new wharf is being installed in salt or brackish water.

In establishing quality specifications, two general areas of concern should be concentrated. The first is the condition and grade of the wood material. Lumber and panels should be of the grade and quality as specified or represented. Poles, piles, and heavy timbers should be inspected prior to treatment to determine if any defects that could cause structural weakening are present and that the material is of the quality needed. The second area of concern is whether the material has been properly treated with the preservative as specified or represented.

Figure 7-1. *Typical lumber grade stamp as approved by ALSC and its interpretation for Douglas-fir lumber. (Courtesy of ALSC)*

KEY:

a - Trademark indicates agency quality supervision.

b - Mill Identification - firm name, brand or assigned mill number.

c - Grade Designation - grade name, number or abbreviation.

d - Species Identification - indicates species individually or in combination.

e - Condition of Seasoning at time of surfacing:

S-Dry - 19% maximum moisture content

MC 15 - 15% maximum moisture content

S-GRN - over 19% moisture content (unseasoned)

KD - kiln-dried to 19% maximum moisture content.

This chapter first discusses what a consumer of small quantities of treated wood products should look for in making a purchase. Then, details are given on what an architect, contractor, specifier, or purchasing agent can do to provide maximum assurance that all standards and specifications have been closely followed and a quality-treated product delivered.

Purchases for Residential and Agricultural Applications

When purchasing small quantities of preservative-treated material, such as lumber or small timbers from local lumber yards, the material should be examined to determine if the grade (appearance) of the lumber is acceptable for the intended end-use. Some yards will carry more than one grade. The grade is normally stamped on each piece (**Fig. 7-1**). Price is usually directly related to grade. Some material will also contain more bark edges or wane than others. Bark edges may be objectionable if appearance is important. For the lumber grade mark to be valid after treatment, the lumber must adhere to the grade requirements and the moisture content of the grade as represented.

Because heartwood does not accept preservative treatment well, its presence in southern pines should be limited in critical members and in areas of high degradation hazard. Some heartwood is nearly always present, but some shipments will have a higher percentage (**Fig. 7-2**). Pieces wider than about 6 inches will also tend to have more heartwood. If possible, those pieces containing heartwood should be placed where exposure to biological agents is less severe. In contrast, Douglas-fir is almost all

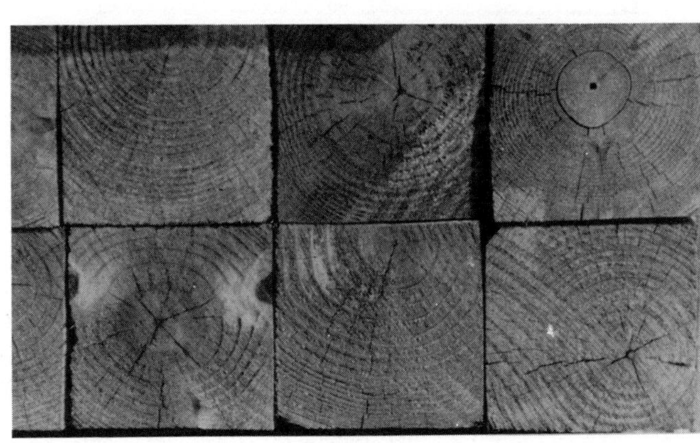

Figure 7-2. *Heartwood in treated Southern Pine lumber.*

heartwood and different treating specifications and preservatives apply.

The next concern is the preservative treatment. Each piece should be stamped with a quality mark (**Fig. 7-3**) or contain a tag (**Fig. 7-4**). The American Lumber Standard Committee (ALSC) accredits agencies using the stamp or tag. This represents the industry's method of monitoring itself and providing consumers at the retail level with a quality product. In essence, the system has established applicable quality control standards and accredited independent inspection agencies to monitor subscribing treating facilities and determine if they are in compliance. This ALSC quality control overview is preferable to simple treating plant certificates or other claims of conformance made by the producer without inspection by an independent agency. If the treated-wood products do not carry the stamp or tag and the supplier is not willing to provide documentation that the appropriate AWPA standards were followed, there is no assurance of the quality of the product. For some applications such as decks, an ink stamp on each piece of lumber is aesthetically objectionable. Therefore, end tags are used by some suppliers.

ABC [1]	19_-19_ [3] GROUND CONTACT [6] .40 [5]
XXX [7]	AWPA____STDS [2] PRESERVATIVE [4] KDAT [8] X-XX [9]

Figure 7-3. *Typical quality mark for lumber preservative treated to conform to the ALSC accreditation program. (Courtesy of ALSC)*

KEY:

1 - Identifying symbol, logo, or name of the accredited agency.
2 - Applicable AWPA commodity standard.
3 - Year of treatment if required by AWPA standard.
4 - Preservative used, which may be abbreviated.
5 - Preservative retention.
6 - Exposure category (e.g., above ground, ground contact).
7 - Plant name and location, or plant name and number, or plant number.
8 - If applicable, moisture content after treatment (KDAT is kiln dried after treatment).
9 - If applicable, length, and/or class.

Purchases for Commercial Installations

To provide maximum assurance that a quality product is being received, it is necessary to (1) determine if the type and quality of the wood material is suitable for the end-use intended and (2) make certain that the wood preservative and the preservative-treatment process are in accordance with applicable standards. Each step of the process must follow set procedures and be appropriately documented. Independent inspection agencies are often retained to assure that the material, preservative, and treatment specifications are strictly followed.

Material Specification

Wood is a biological material and is naturally variable. The variability is related to a large number of factors including species, growth rate, age, genetic makeup, mechanical damage such as cross breaks, and natural defects such as knots and shakes. Some defects, such as cross breaks in poles and piles, can be detected only after the bark has been removed and before the product is treated. These products are often inspected before preservative treatment. The industry refers to this as inspection "in the white." Existing material specifications have been developed that group (or grade) wood products

Figure 7-4. *Waterborne preservative treated lumber showing end tags.*

with known structural properties and engineering design values.

For complete material standard specifications for different products, the following sources are referenced:

Poles: American National Standard Specifications and Dimensions for Wood Poles, ANSI 05.1 1992. American National Standards Institute, Inc., 1430 Broadway, New York, NY 10018. (212) 642-4900.

Piles: American Society for Testing and Materials, ASTM D25-91, Standard Specifications for Round Timber Piles. American Society for Testing and Materials, 1916 Race Street, Philadelphia, PA 19103. (215) 299-5400.

Lumber and Timbers: U.S. Product Standard PS 20-94 for American Softwood Lumber Standard. National Bureau of Standards (currently National Institute of Standards and Technology), U.S. Dept. of Commerce, Gaithersburg, MD 20899. (Available from American Lumber Standard Committee, Box 210, Germantown, MD 20875-0210. (301) 972-1700.

Structural Use Panels: Performance Standards and Policies for Structural-Use Panels. Feb., 1991. American Plywood Association, Box 11700, Tacoma, WA 98411. (206) 565-6600.

Treatment Specifications

The proper choice and application of preservative for the intended end-use are critical to the successful performance of the treated product. The AWPA standards require different preservative retention levels for different use environments. Substitution of materials properly treated for one environment into another environment may contribute to early failure or unacceptable aesthetic qualities. For example, marine exposure requires a much heavier retention of preservative than is required for use in ground contact. Marine-exposed wood may also utilize preservatives such as creosote that are not acceptable for indoor use. Recommended retention levels and applicable AWPA standards by product and preservative for southern, ponderosa, and red pines are summarized in **Table 7-1**; Douglas-fir, western hemlock, and the species group Hem-Fir are summarized in **Table 7-2**. For complete details, users should refer to the actual standards. Note that some preservatives are restricted to certain species. Also, other less common species are given in AWPA standards. These standards address chemicals, process requirements, and performance requirements for preservative pressure-treated wood products. Acceptable preservatives and carriers are identified and defined. The standards also define the amount of each type of preservative and the depth of penetration required by commodity and species. Processing parameters, such as temperature, are also defined. For some commodities, such as poles, the preservative penetration and retention within a given assay zone will vary with species. Sampling protocols for inspecting treated wood products and methods for care and handling products after treatment are also addressed.

Although AWPA standards are generally accepted by treated-wood producers and purchasers, all wood is not necessarily treated according to these specifications. If AWPA standards are not followed, purchasers should be certain that they understand the technical aspects of wood preservation and that the specifications as written will provide a product meeting or exceeding long-term performance expectations. Inspection must be conducted to determine if the specifications are being carried out.

Treatment to refusal is a term still used in standards for pressure treatment of a few difficult-to-treat woods such as white oak. When this terminology is used in a procurement specification for pressure-treated products, it does not require a minimum amount of preservative (retention) or a minimum depth of treatment (penetration). Whenever the minimum retention and penetration are listed in standards for pressure-treated wood products, this terminology, treatment to refusal, should not be used. Instead, cite the standard as published. Use of treatment to refusal could allow acceptance of poorly treated products that would not otherwise meet standard requirements.

The NWWDA specifies acceptable preservatives and the degree of water repellency required for nonpressure treatments.[20] The AWPA is currently working on standards for nonpressure treatment.

Additional specifications on wood preservation by pressure treatment and water repellents have been published by the U.S. General Services Administration. These specifications have not been updated for several years, and it is more appropriate to reference current industry standards.

Inspection

Procedures

Inspection to determine if the material, the preservative, and the treatment process comply with applicable standards and procurement contracts can be done at several different times and locations. Pretreatment inspections are made of materials after the bark has been removed, but before treatment. These inspections, especially for large members such as poles and piling, are normally done at the treating plant just before treatment.

Whether the treating plant complies with processing parameters as specified in procurement contracts can only be determined at the plant. For example, gauges and charts are examined during and after treatment and a sample of the preservative is taken to verify that the treatment process is being performed in compliance with requirements of the procurement contract.

The material can also be inspected at the destination regardless of whether it was inspected at the plant or not. However, for large products such as poles, piles, and

Table 7-1. *Recommended retention levels and applicable AWPA standards*[a] *for southern, ponderosa, and red pine by product and preservative based on AWPA standards.*[24]

Use	AWPA standard(s)	Retention assay of treated wood (lb/ft^3) ACQ[b]	ACZA	CCA	Creosote	Creosote petroleum[c]	Creosote solutions[c,d]	Pentachlorophenol[e]	Copper naphthenate[b]
LUMBER, TIMBER, AND PLYWOOD									
Above Ground	C2/C9	.25	.25	.25	8	8	8	.40	.04
Permanent Wood Foundation[f]	C22	NR[g]	0.60	0.60	NR	NR	NR	NR	NR
Salt Water	C2/C9	2.5	2.5	2.5	25	NR	25	NR	NR
Soil & Fresh Water	C2/C9	0.40	0.40	0.40	10	10	10	0.50	0.06
PILES									
Foundation, Land or Fresh Water	C3	NR	0.80	.80	12	12	12	0.60	0.10
Salt Water									
Prevalent Marine Organism:									
Limnoria tripunctata only	C18	NR	2.5[h] and 1.5	2.5[h] and 1.5	NR	NR	NR	NR	NR
Pholads only	C18	NR	NR	NR	20	NR	20	NR	NR
Sphaeroma terebrans or for both pholads and *Limnoria tripunctata* use a dual treatment									
First treatment	C18	NR	1.0	1.0	-	-	-	-	NR
Second treatment	C18	NR	-	-	20	-	20	-	NR
Teredo only	C18	NR	2.5[h] and 1.5	2.5[h] and 1.5	20	NR	20	NR	NR
POLES									
Building									
Round and Sawn Timber	C23/C24	NR	0.60	0.60	9.0	NR	NR	0.45	NR
Utility—Service Conditions:									
Normal	C4	NR	0.60	0.60	7.5	7.5	7.5	0.38	0.08
Severe (high incidence of decay and termite attack)	C4	NR	0.60	0.60	9.0	9.0	9.0	0.45	0.13

timbers, specialized handling equipment may be required.

Although some inspection programs require examination of every piece of treated material, most inspections are done by sampling. Sampling is used primarily to reduce inspection costs. The frequency and intensity of sampling is often defined in treating standards, but the purchaser has the option of requiring more rigorous sampling. Although commercial wood-treating companies maintain some form of quality control through internal inspections, additional independent inspection at the plant site and destination reduce the risk of accepting nonconforming items.

Inspection of wood products before, during, and after treatment requires experienced individuals with specialized knowledge and equipment. Increment borings may be made to determine preservative retention and penetration. Therefore, the process is normally completed by an independent agency accredited by ALSC (a current list is available from the committee.) Some large routine buyers, such as utility companies, may employ their own inspectors, but most purchasers employ independent inspectors.

Table 7-1. *Recommended retention levels and applicable AWPA standards[a] for southern, ponderosa, and red pine by product and preservative based on AWPA standards.[24] (continued)*

		Retention assay of treated wood (lb/ft^3)							
Use	AWPA standard(s)	ACQ[b]	ACZA	CCA	Creosote	Creosote petroleum[c]	Creosote solutions[c,d]	Pentachlorophenol[e]	Copper naphthenate[b]
POSTS									
Commercial-Residential									
Fence:									
Round, half-, and quarter-round	C5	0.40	0.40	0.40	8	8	8	0.40	0.055
Sawn four sides	C2	0.40	0.40	0.40	10	10	10	0.50	0.06
Highway Construction									
Fence, Guide, Sign, and Sight Posts:									
Round, half-, and quarter-round	C14	NR	0.40	0.40	8	8	8	0.40	0.055
Sawn four sides	C14	NR	0.50	0.50	10	10	10	0.50	0.06
Guardrail and Spacer Blocks:									
Round	C14	NR	0.50	0.50	10	10	10	0.50	0.069
Sawn four sides	C14	NR	0.60	0.60	12	12	12	0.60	0.072

[a] AWPA standards detail plant operating procedures for pressure treatment of wood. These standards include minimum vacuum, pressure, penetration requirements, maximum steaming, and temperature allowances. AWPA also details retention and assay zone requirements for each commodity, preservative and wood species. AWPA standards make it unnecessary for specifications to include detailed requirements on penetration and allowable processes. Generally, it is desirable to specify the preservative desired, the intended application, necessary retention and to reference appropriate AWPA standards. AWPA standard C-1 applies to each of the treating processes and all types of material.

[b] These preservatives have only recently been added to the AWPA Standards. Hence, the eight wood species and commodities approved for treatment with them are limited and subject to change. Copper naphthenate retention levels are expressed as copper metal.

[c] Not recommended where cleanliness and freedom from odor are necessary.

[d] When these preservatives are specified, the creosote shall conform to AWPA Standard P1 and the creosote solutions shall conform to Standard P2.

[e] Pentachlorophenol in suitable solvents or waterborne preservatives can provide a clean paintable, odorless, dry surface. When one or more of these features are required, the processor should be advised when the order is placed.

[f] When the lumber and plywood parts have been treated in accordance with AWPA C-22 standard and the foundation is fabricated and constructed in accordance with National Forest Products Association Technical Report No.7, the foundation should have a life equal to or greater than the rest of the structure.

[g] NR — Not Recommended.

[h] Assay retentions are based on two assay zones: 0 to 0.50 in. and 0.50 to 2.0 in.

Documentation

All structural material including treated piling, building poles, lumber, timbers and plywood should have a mark or stamp on each piece indicating that it was inspected before treatment. Brands, stamps, or tags indicating that the product was inspected after treatment and that it complies with the appropriate AWPA standards should also be present. Common symbols for wood species and preservatives that may be used on all commodities treated in accordance with AWPA specifications are given in **Table 7-3**. An inspection report or certificate of an independent inspection agency certifies that the items comply with the appropriate AWPA standards. Reports or certificates should show actual test results for penetration and retention of preservative. Certificates from the treating plants are of dubious value. Millwork, inspected and approved by NWWDA, should bear a

Table 7-2. *Recommended retention levels and applicable AWPA standards[a] for western species by product and preservative based on AWPA standards.*[26]

		Retention assay of treated wood (lb/ft³)				
Use	AWPA standard	ACZA	CCA[b]	Creosote	Pentachlorophenol	Copper naphthenate[a]
GLUED LAMINATED BEAMS (treatment after lamination)						
Above Ground	C28	NL	NL	8	0.40	NL
Ground or Fresh Water Contact	C28	NL	NL	10	0.50	NL
LUMBER, TIMBER, AND PLYWOOD						
Above Ground:						
Lumber, Timber	C2	0.25	0.25	8	0.40	0.04
Plywood	C9	0.25	0.25	8	0.40	NL[c]
Bridges, Important Structural Members	C2/C14	0.60	0.60	12	0.60	0.075
Ground Contact or Fresh Water:						
Lumber, Timber	C2	0.40	0.40	10	0.50	0.04
Plywood	C9	0.40	0.40	10	0.50	NL
Marine						
In Salt Water:						
Dual treatment	C2/C18	1.50	1.50	20	NR	NL
Single treatment	C2/C18	2.50	2.50	25	NR	NL
Out of Water:						
Not subject to splash	C2/C18	0.40	0.40	10	0.50	NL
Subject to splash	C2/C18	0.60	0.60	12	0.60	NL
Permanent Wood Foundations (kiln-dried after treatment):						
Lumber	C22	0.60	0.60	NR[d]	NR	NL
Plywood	C22	0.60	0.60	NR	NR	NL
PILES						
Foundation:						
Round	C3	.80-1.00	.80-1.00	12-17	.60-.85	NL
Sawn	C24	.80	NR	12	.60	NL
Freshwater	C3	.80-1.00	.80-1.00	12-17	.60-.85	NL
Salt Water:						
Borer Hazard, moderate[f]	C18	2.5	2.5	20[g]	NR	NL
Borer Hazard, severe[f]	C18	2.5	2.5	NR	NR	NL
Dual treatment[f,h]	C18	1.0	1.0	20[g]	NR	NL

"Quality Certified" brand (**Fig. 7-4**) as proof of inspection.

Spot Checking Different Commodities

Poles. There are two types of poles: round building and power poles. To check for evidence of inspection prior to treatment, look on the tip (top) end of the poles for a brand that identifies the inspection agency and the inspector. Upon inspection "in the white," poles meeting material standards are hammer stamped on the tip end. This is the only stamp that is ever put on the tip end.

Table 7-2. *Recommended retention levels and applicable AWPA standards*[a] *for western species by product and preservative based on AWPA standards.*[26] *(continued)*

Use	AWPA standard	Retention assay of treated wood (lbs/ft^3) ACZA	CCA[b]	Creosote	Pentachloro-phenol	Copper naph-thenate[a]
POLES						
Agricultural:						
Round[e]	C16	0.60	0.60	7.5-16	0.38-.60	NL
Sawn[e]	C16	0.60	0.60	12	0.60	0.075
Construction:						
Round[e]	C23	0.60	0.60	12	0.60	NL
Sawn[e]	C24	0.80	0.80	12	0.60	NL
Utility:						
Electric/Telephone/Lighting	C4	0.60	0.60	9-16	0.30-0.80	0.075-0.150
POSTS						
Agricultural:						
Round	C16	.60	.60	7.5-16	.38-.60	.055
Sawn	C16	.60	.60	10-12	.50-.60	.060
Construction:						
Round	C5	.40	.40	6	.30	.055
Sawn	C15	.40	.40	10	.50	NL
Guard Rail (including blocks):						
Round	C14	.50	.50	10	.50	.069
Sawn four sides	C14	.60[i]	.60	12	.60	.072

[a] American Wood Preservers Association (AWPA) specifications are the principal wood-treating standards used throughout the United States. By assay procedure—pounds per cubic foot. Retention varies with species. See referenced AWPA Standard. Copper naphthenate retention levels are expressed as copper metal.

[b] It is generally recognized that Douglas-fir is extremely difficult to treat with CCA to penetration and retention requirements, even when incised. Coastal Douglas-fir, from a few geographical areas, has been found suitable for treatment with CCA. Douglas-fir treated with CCA is not recommended for Permanent Wood Foundations.

[c] Not listed in AWPA standards.

[d] Not recommended in AWPA standards.

[e] Building Poles—special uses—to be used where replacement of poles would be difficult and where exceptional durability is required. Involves requirement for very deep penetration, i.e., ½ the radius to 2-½ in.

[f] On the Pacific Coast north of San Francisco where *Teredo* attack is expected and where *Limnoria tripunctata* is not expected, creosote or creosote-coal tar solutions will provide adequate protection. On the Pacific Coast, San Francisco and south where *Teredo* and *Limnoria tripunctata* are expected and where pholad attack is not expected, either dual treatment, or high retentions of ACZA or CCA, will provide adequate protection. In tropical waters where *Limnoria tripunctata* and wood-boring pholad attack are expected, such as Hawaii, the Caribbean, or off the Mexican coastline, dual treatment provides the maximum protection known at present. On the Eastern seaboard, the line between moderate and severe borer hazard is usually drawn at Norfolk, Virginia.

[g] Coastal salt water environments. Creosote for such use shall conform to the current AWPA Standard P-13.

[h] Dual Treatment for marine use involves two separate preservatives: a waterborne preservative followed by a creosote treatment.

[i] State specifications may differ, allowing 0.40 retention.

Check the brand applied by the supplier to determine the type of treatment and preservative retention level. Poles are branded on the face side of the pole 10 feet from the extreme butt for poles 50 feet or less and 14 feet from the extreme butt for poles more than 50 feet long. These brands contain the following information:

(1) Name of supplier
(2) Plant designation
(3) Year of treatment
(4) Species of timber and preservative treatment
(5) Preservative retention level
(6) Class and length

Power poles are often treated in accordance with the current AWPA Standard C4 and may be inspected under Power Pole Standard REA DT-5C. To determine that power poles have been inspected, check for a hammer mark of an approved inspector/agency on the butt end of each pole.

Table 7-3. *Brands used on wood products treated in accordance with AWPA standards.*

Typical brand and key[1]

ABCO	Supplier's brand
D	Plant designation
60	Year of treatment[2]
SPC	Species of timber and preservative treatment
—	Retention[3]
7-30	Class and length

Species

AS	Ashes
BE	Beech
BI	Birches
BW	Black Walnut
DF	Douglas-fir
ED	Northern White (Eastern) Cedar
ES	Engelmann Spruce
GU	Gum
HI	Hickories
JP	Jack Pine
LO	Locust
LP	Lodgepole Pine
MA	Maples
NP	Red Pine
OA	Oaks
RW	Redwood
SP	Southern Pine
SS	Sitka Spruce
WC	Western Red Cedar
WF	Western Firs
WH	Western Hemlock
WL	Western Larch (Western Tamarack)
WP	Ponderosa Pine
WS	White Spruce
YC	Alaska Yellow Cedar

Preservatives[4]

Organic Preservatives

C	Creosote
CM	Creosote for marine use
PB	Pentachlorophenol in Volatile Petroleum Solvent (LPG)
PC	Pentachlorophenol in Light Hydrocarbon Solvent
PD	Pentachlorophenol in Chlorinated Hydrocarbon Solvent
PA	Pentachlorophenol in petroleum
XA 80/20	Creosote-Petroleum Solution
XB 70/30	Creosote-Petroleum Solution
XC 60/40	Creosote-Petroleum Solution
XD 50/50	Creosote-Petroleum Solution

Note: Double treatment, salt and creosote, letter "C" to be placed ahead of salt symbol. Where organic preservative is other than creosote, use the symbol for that preservative instead of the symbol for creosote.

Organometallic Preservatives

B	Copper Naphthenate in Creosote
N	Copper Naphthenate in Petroleum
C8	Copper-8-Quinolinolate
TO	Tributyltin Oxide

Inorganic Preservatives

SA	Acid Copper Chromate (ACC)
SB	Ammoniacal Copper Arsenate (ACA)
SC	Chromated Copper Arsenate (CCA-Type A)
SE	Chromated Zinc Chloride (CZC)
SF	Copperized Chromated Zinc Arsenate (CuCZA)
SJ	Chromated Copper Arsenate (CCA-Type B)
SK	Chromated Copper Arsenate (CCA-Type C)
SZ	Ammoniacal Copper Zinc Arsenate (ACZA)

[1] This information may be placed in the material in some other form and/or order if authorized by the purchaser.
[2] May also include month as "10-60".
[3] If retention is determined by assay, place letter "A" after numeral.
[4] If desired, full names, or the abbreviations shown, or trade names may be used for preservatives instead of the recommended standard symbols.

Building poles are treated in accordance with the current AWPA Standard C23. In shipments under the ALSC accreditation program, look for the quality mark of the accredited agency as evidence that building poles were inspected.

Piles. There are three types of piles: marine, general (usually called land and fresh water), and foundation. Marine piles are exposed in salt or brackish water; general piles are used in fresh water or in soil, but not capped with concrete. Foundation piles are entirely embedded in the ground and capped with concrete.

To determine if piling was inspected prior to treatment, look for the inspector's hammer brand on the tip end of the piling just as with poles.

To determine the type of preservative that was used and the retention to which the piling was treated, check the brand 5 and 10 feet from the butt end of the pile. These brands should show the following:

(1) Name of supplier
(2) Plant designation
(3) Year of treatment
(4) Species of timber and preservative treatment
(5) Preservative retention level
(6) Length

The symbols used for preservatives and wood species are shown in **Table 7-3**. The treatment of piles is addressed in the current AWPA commodity Standards C3, C14, and C18.

For those piles procured with the services of the inspection program offered by ALSC, check the piles for the quality mark of an ALSC accredited agency. To determine the intended use of piling inspected under the ALSC program, read the letter symbols for the respective AWPA standard (**Tables 7-1** and **7-2**).

Lumber, Timber, and Plywood. Lumber, timber, and plywood should be graded prior to treatment, if it is to be used structurally, and stamped to certify that it meets specific requirements for the grade represented. As mentioned previously, each piece of pressure-treated lumber, timber, or plywood should bear a quality mark or stamp that will evidence independent inspection of the wood after it was treated. When checking materials that have been inspected by agencies approved by ALSC, it is important to recognize that there are a number of stamps or marks for lumber, timber, and plywood. The stamp or mark provides information on both the preservative used and the intended use environment for the treated wood. By careful reading of these marks (**Fig. 7-3** and **Tables 7-1** and **7-2**), the person receiving the shipment can determine whether the materials received are appropriate for the planned use.

Figure 7-5. *"Quality Certified" brand that indicates compliance with water-repellent preservative treatment standards of NWWDA.*

Current AWPA Standards C1 and C2 address the general aspects of treating lumber and timbers. Other, more focused standards, address specific commodities. Plywood is covered by AWPA Standard C9. The requirements for treated materials used specifically in marine environments are addressed in AWPA Standards C2, C3, and C18.

Chapter 8
Post-Construction Inspections and Remedial Treatment

Page

Post-Construction Inspection and Remedial Treatment

Because biological deterioration can occur in treated wood, it is important, particularly in large commercial applications, to develop a preventative maintenance program. The program should provide for the inspection of the wood product for biological damage and remedial treatment where appropriate. A determination of the structural reliability of the product in the damaged condition should also be made. In this section, typical deterioration problems are discussed, inspection procedures and equipment outlined, and possible remedial treatments presented.

To understand where serious biological deterioration of treated-wood products can occur, the manner in which these products are protected should be reviewed. The preservative is forced from the outside in. The penetration achieved is dependent on several factors. Seldom are wood products completely penetrated, with the exception of sapwood lumber of permeable species under about 2 in. thick. Thus, there is a shell of treated or protected wood and a core of untreated wood. If the shell becomes broken by seasoning (drying) checks, machining such as sawing or boring, or mechanical damage, the core is subject to biological deterioration.

If the retention of the preservative is not adequate, the treated exterior wood may also begin to deteriorate. A poorly formulated preservative or selection of the wrong preservative for a particular application may also result in deterioration. In these cases, deterioration begins on the outside of the product.

Typical Problems

Lumber and Plywood

Lumber and plywood properly treated for the intended end-use generally do not present a serious deterioration problem. As mentioned, the thickness dimension of standard lumber allows good penetration and as such good performance. Most treated lumber and plywood are also used in above-ground applications such as decks and sill plates, where the hazard level is not as high as when the wood product is exposed to soil or marine borers. Care should be taken to avoid exposure of Southern Pine lumber and timber heartwood faces to marine borers, because of the generally poor preservative penetration.

Posts

Posts, whether round or square, are larger in dimension than standard lumber. They are also often in soil contact, which provides optimum conditions for wood decay and termites. Round posts that have a high percentage of sapwood should treat easily and give good service. Square or sawn posts that have a large percentage of heartwood will be difficult to treat and will likely not perform as well.

Peeler cores are what remains after the outer portion of a larger log is removed for veneer. Cores are no longer included in the AWPA standards for posts. However, cores may still be available, particularly in the form of landscape timbers. Whether round or with two or more flat faces, these cores are mostly heartwood and are usually poorly treated.

Poles

Poles are like large posts and are usually round, but sometimes sawn to rectangular dimensions. When degradation occurs, it is usually in the form of internal decay and insect attack or by external surface decay, usually at or below the groundline. Internal decay can occur as a result of fungus infestations that start in large poles that are airdried in yards before treatment. Large Douglas-fir poles are often airdried for up to 2 years prior to treatment. Decay fungi may become established in the poles and if not heat sterilized during preservative treatment, the fungi will continue to grow and rot the center of the pole after it has been put in service.

If the outer protective shell of a pole is broken after treatment, wood-decay fungi and insects, such as termites and carpenter ants, can gain entry and deteriorate the untreated portion of the pole. The shell may be broken by mechanical damage during handling, deep season checks (**Fig. 8-1**) developing after treatment, woodpecker damage, or other actions.

As a result of the constant presence of moisture, internal decay is most common at the groundline. Internal decay will also develop in pole tops cut or bored in the field when supplementary treatment is neglected. Butt ends on poles should NOT be cut when they are set in or on the ground. This exposes a central core of untreated wood at the bottom of the pole and provides easy access for termites and decay fungi.

External decay in poles is most common at or below the groundline (**Fig. 8-2**). External decay of relatively new poles usually reflects improper application of the preservative or use of treating solutions with poor preservative properties. As poles age, external decay may develop as the effectiveness of the treatment begins to decline.

Termites and carpenter ants can infest the internal untreated portion of poles with little external visual evidence of their presence. Some termite galleries may be present if the insects are trying to bridge over treated wood. In addition, if a carpenter ant infestation has occurred, scattered bits of very fibrous and sawdust-like frass may be present in the area. Because a break in the protective shell must occur before these insects can reach and infest the untreated wood, decay is also likely to be present.

Damage from vertebrate organisms, such as woodpeckers, is usually apparent. Binoculars should be used when inspecting large poles. If the damage is fresh, broken pieces of wood from the excavated hole should be present on the ground. Decay will often be associated with older damage (**Fig. 8-3**).

Piles

In many respects, piles are much like poles, and those portions of piles exposed above groundline to the atmosphere or fresh water are subject to the same problems as poles. However, piling is subject to additional problems.

Figure 8-1. *If deep seasoning checks occur in poles after preservative treatment, the interior is subject to decay. This roofing hatchet is buried about 5 in. in the pole, and it is still not at the bottom of the check.*

Because it is difficult to drive piles to a uniform depth, it is common practice to cut off the top end. This practice exposes untreated wood to fungal attack and a hollow core often results (**Fig. 8-4**).

Marine borers are one of the worst enemies of piling driven in salt or brackish water. *Limnoria* and *Sphaeroma* often attack piling at the intertidal zone and their presence is then obvious (**Figs. 3-22, 3-23, 3-24**). Shipworms generally attack wood below the tidal zone, and they are not conspicuous.

Timbers

Timbers are sawn members 5 or more nominal inches in their least dimension. They are also called beams,

Figure 8-2. *External or surface decay as seen in this pole after removal from service is most common at the groundline.*

Figure 8-3. *Serious internal decay has occurred in this pole because fungi gained access from the woodpecker damage and a small bore hole (upper left).*

stringers, posts, or girders. Because timbers are sawed products, the larger dimensioned ones, at least, usually do not contain a continuous, easily treated sapwood band. Southern Pine may be an exception in that it contains a high percentage of sapwood, but the trees required to produce large timbers will have a substantial amount of heartwood. Small timbers cut from the core of medium-sized trees will also have a substantial component of heartwood. Therefore, incising of most large timbers should be used to help improve the preservative treatment.

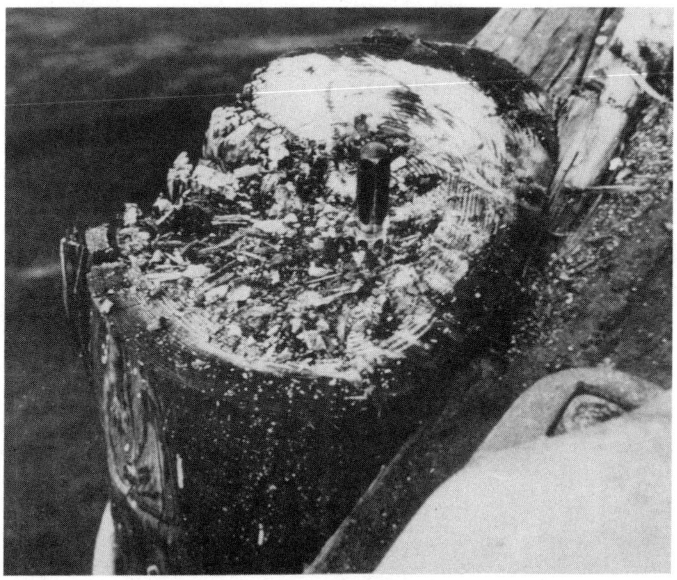

Figure 8-4. *Because the end of this pile was cut off after driving, untreated wood was exposed and decay resulted.*

Figure 8-5. *This timber contains a pith center that resulted in a deep radial seasoning check.*

Many timbers will contain a pith center that results in a deep radial seasoning check from the surface to the center (**Fig. 8-5**). This check often penetrates past the treated zone and can permit entry of decay and insects if it occurs after treatment. For this reason, it is strongly recommended that timbers be dried prior to treatment to a lower moisture content than they are likely to reach when in service. However, such drying is difficult and can often be costly.

Heavy timber construction often requires substantial shaping, boring, and other machining at the erection site. This machining can break the protective shell. Ideally, engineering drawings should be consulted so that boring and other machining may be done prior to preservative treatment. The shell can also be broken by nailing, which often occurs in timber construction. As a result, interior decay can become a problem. **Figure 8-6** shows the likely locations where decay might occur.

Inspections

Post-construction inspections are carried out to determine if any biological damage has occurred and if remedial treatment or replacement should be considered. Obviously, if some damage has occurred, the structural integrity must be carefully considered. Advanced biologi-

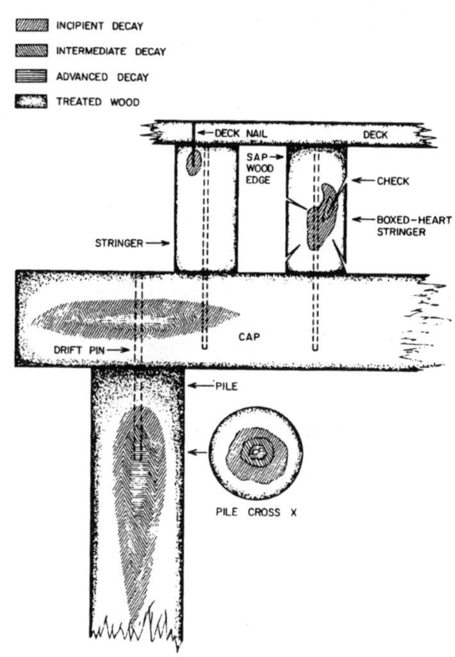

Figure 8-6. *Schematic of typical wood bridge construction showing likely locations where decay could develop.*

cal damage is sometimes obvious. Unfortunately, damage often occurs in the interior of large timber products at or below groundline or under water and is difficult to detect. Furthermore, even in the early stages when decay can easily go undetected, serious loss of strength can occur. Therefore, it is best to have someone with technical training, experience, and proper equipment carry out inspection programs.

This section overviews inspection procedures for poles, piles, and timbers. Because treated structures are intended to endure for many decades, careful records of each inspection and recommended actions should be made. Photographs can be particularly useful. These materials are useful in documenting the rate of deterioration over time and providing a valuable training aid for subsequent inspectors.

Visual

A visual inspection represents the very basic level and in many cases it can yield valuable information. In a visual inspection, machined, damaged areas, and checks should be carefully examined. The size and location of seasoning checks should be noted. In general, the deeper the check, the more likely untreated wood is exposed.

External evidence of advanced decay includes features such as fruiting bodies, abnormal surface shrinkage such as localized depressions or sunken faces on wood surfaces, loose joints, abnormal deflections, crushing, cracking, and insect activity. Carpenter ants and termites prefer damp wood.

Other visual evidence such as water marks from periodic wetting; rust stains, especially if from wood-penetrating fasteners or hardware; and plant growth such as moss or other vegetation, especially in cracks or crevices indicates that existing conditions are, or have been, conducive to decay.

Remember, only decay in the advanced stages is readily apparent. Incipient decay can extend 4 feet or more above internal rotten areas in Douglas-fir poles. The presence of fungi in wood where decay has not progressed appreciably can be detected only by culturing or microscopic examination of the wood.

Surface decay is detected by the pick test or probing. The pick test (**Fig. 3-2**) will detect infected wood where strength loss of at least 10 percent has occurred. Scraping using a shovel or triangular blade is often used to detect below ground decay on the pole surface. The decayed wood is simply scraped or cut away until sound wood is evident. In poles, surface decay usually occurs at or below the groundline so digging is generally necessary to detect decay.

Sounding

Sounding is probably the most common method used by experienced individuals for detecting internal voids in large wood members. The wood member is firmly hit with about a 1-lb hammer. A crisp sound usually indicates soundness. A dull thud indicates wet and possible rotten wood, and a "drum" sound indicates a hollow or void area. The sound produced varies considerably, but experience will eventually lead to distinguishing between sound wood and the various defects. Naturally, to develop experience, it helps to sound a member and then bore or cut it apart to determine the defects that are present.

Boring

Boring is generally done where decay or insect attack is suspected and it needs to be confirmed. Increment borers are most commonly used (**Fig. 8-7**). The core can be closely examined at the site and also saved for later culturing or microscopic examination. Wood members that sound suspicious should be bored near deep checks. Poles should generally be bored at 1 foot below the groundline. If rot is detected, borings should be made at three or four more points around the circumference to determine its extent. The shell thickness, depth of preservative treatment, and pole circumference are determined. Minimum circumference tables can be used to determine if the pole should be replaced, reinforced, field treated, or scheduled for reinspection.

When boring holes above ground, the tool should be oriented slightly upward. This prevents water from accumulating in the hole afterwards. At groundline, a 45°

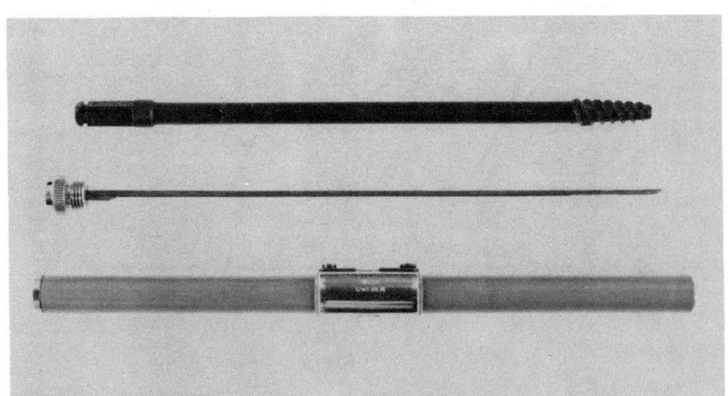

Figure 8-7. *Increment borer consisting of three parts: extractor, hollow bit, and case that also serves as a handle for turning coring bit into and out of wood members.*

angle downward is generally used. All openings made during inspection should be treated with the appropriate preservative. The holes should be plugged with preservative treated dowels. These dowels can loosen and work out of the holes so treatment of the hole is important. Protective goggles and other safety equipment should be worn as preservative can squirt out of the hole as the dowel is being driven or as the hammer strikes the treated wood.

Moisture Meters

Resistance-type moisture meters (**Fig. 8-8**) can detect the moisture content of wood to a depth of about 2-½ in. Because the high moisture content of decaying wood greater than 25 percent causes a steeper than normal moisture gradient, the meter is often useful for determining the extent of decay in poles and other timbers. Meter readings greater than 20 percent and steep moisture gradients indicate decayed wood in Douglas-fir poles. Moisture readings less than 20 percent indicate the absence of active decay to the depth of the electrode.

The batteries and calibration of the meter should be checked frequently. For the meter to read from the tips of the electrodes, the coating on the electrode shanks must be intact. When necessary, meter adjustments for ambient temperatures and wood species must be made. Oilborne preservatives normally do not affect meter readings, but waterborne preservatives may cause erroneously high readings, particularly at moisture content levels greater than 25 percent. Salt absorbed from sea water will result in abnormally high readings. Comparative tests for treated and untreated wood should be conducted to determine the effect of preservatives on the meter reading.

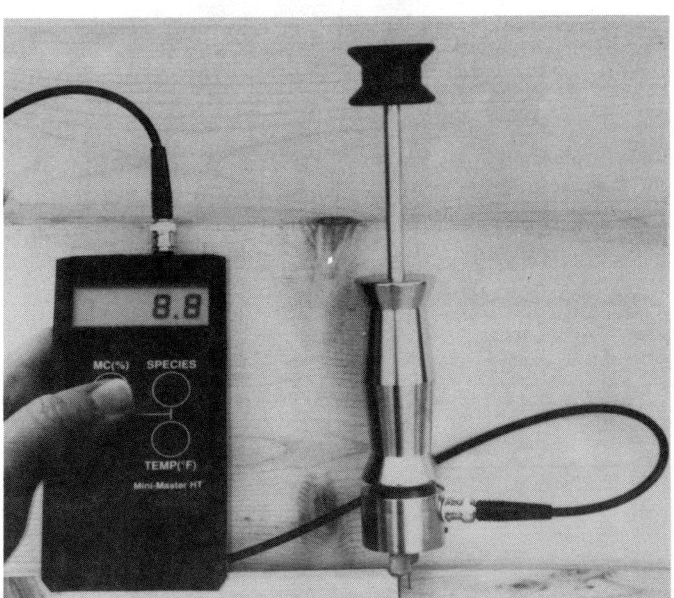

Figure 8-8. *Moisture meter suitable for inspection work. (Photograph courtesy of Lignomat USA Ltd.)*

Sonic Testing

Some sonic testing equipment is currently available and other nondestructive equipment to estimate residual strength in standing poles is being developed. This technical equipment is generally employed only by technical experts specializing in its use.

Culturing and Identification

Because early decay can affect the strength properties of the wood, it is important that the presence of decay fungi be detected and treated before visual damage is present. Culturing is among the most reliable means for detecting early stages of decay. It is done by collecting cores or shavings in the field. When coring, the core is placed in a plastic straw, labeled, and the ends of the straw stapled shut. The cores are brought to the laboratory and culturing begins within 24 hours. The cores are observed frequently for fungal growth. Microscopic examinations are made at the end of 3 to 4 weeks to determine if the fungi are capable of causing decay.

It is usually beneficial to identify insects if an infestation has occurred. If field identification is not possible, either the insects, their frass, or a portion of the wood with typical damage should be collected. University Extension programs often provide identification services.

Marine borers are unique to piling and deserve special consideration; their characteristics are discussed in Chapter 3. The presence of water serves as a physical barrier to inspection. In some cases, borers are present in a salt water wedge located under incoming fresh water.

Limnoria can be detected during periods of low tide and inspectors normally wade out to check piling in shallow water and use small boats or floats in deeper water. Damage signs include the hourglass shape of piles in the tidal zone, numerous bore holes in the wood, and a general softening in the attacked areas. Holes made by hooks and tongs during construction or use become oval-shaped as attack progresses, ends of cutoff bracing are hollowed out, and bolts and bracing are loosened. *Limnoria* also attack wood inside of holes, cracks, or gouges. Therefore, a flashlight or head lamp is useful. A scraper or probe is used to remove fouling organisms and check inside holes and between adjourning members.

Shipworms are much more difficult to detect because they leave only tiny holes on the wood surface and concentrate their attack within the wood (**Fig. 3-22**). Observant divers in clear water may be able to see the nearly transparent siphons or the tiny pallets protruding from the pile surface, if the shipworms are alive. The extent of internal damage cannot be assessed visually.

Sounding of piling by divers can provide some information about the degree and location of borer damage. However, the procedure is highly subjective. Increment boring can also be used. However, with poor underwater visibility and fouling, it is difficult to determine where to bore and the odds of hitting a destroyed area are reduced due to the size of the bore. The bore holes must be plugged.

The potential damage from marine borers can be assessed by immersing "sacrificial" untreated wood samples at the site in question. The simplest way to do this is to band or nail (use galvanized nails) lengths of untreated 2 by 4's, posts, or similarly sized wood pieces to vertical posts or piling already installed in the area of interest (**Fig. 8-9**). These samples are removed at 1 month intervals and examined for damage. For shipworms, the blocks must be split open to accurately detect attack. Where freshwater flow is present, some blocks should be submerged and held in the salt water wedge located under the freshwater.

A sonic testing method that locates and quantitatively assesses marine borer and other internal damage for in-place timber piles has been developed.[1] Again, this specialized equipment is used only by technical experts.

Remedial Treatments

The remedial preservative treatment of wood in existing structures is a common practice. In many cases, only a small percentage of the wood members need treatment. The replacement value of many treated-wood structures or even individual members is substantial and replacement often results in interrupted service. Contractors who specialize in inspection work and remedial treatment are available.

Liquid Preservatives

Liquid preservatives with an oil base, such as pentachlorophenol or copper naphthenate, are sometimes used in remedial treatment. At the time of treatment, the wood should be as dry as possible. The surface is completely flooded and allowed to absorb as much preservative as possible. This is a superficial treatment and its effectiveness will be limited accordingly. The newer water soluble borates diffuse into wet wood; they are being used where the treated wood is not subject to wetting.

Liquid preservatives are also used for the treatment of internal voids. They are most successful on cedar and other woods that develop well-defined rot pockets and the transition from rotten to sound wood is abrupt. Liquid preservatives are least effective on Douglas-fir that has poorly defined rot pockets. To arrest internal decay, water-soluble chemicals are forced into the voids and diffuse through the wet wood. Application of liquid preservatives without pressure can also be effective, but penetration is primarily along the grain of the wood.

Bandage or Groundline Treatments

Bandage or groundline treatments have long been used as remedial treatments to control decay at the groundline of utility poles. Treatments usually consist of a blend of preservatives with differing capacities to diffuse into the wood. This mix of chemicals provides protection on the surface and deep within the pole. These preserv-

Figure 8-9. *Untreated and lightly treated 2 by 4's fastened to an existing marine piling provide a simple means for determining which marine organisms are present.*

atives are brushed on or injected 2-½ in. into the wood with a hollow needle. The outer wrap, often black polyethylene film with a paper back, should be durable and extend from slightly above to 18 in. or more below the groundline. The layers are overlapped, and the wrap is fastened tightly to the pole to shed water.

Fumigants

Fumigants can control internal decay for up to 20 years in pressure-treated Douglas-fir transmission poles and bulkhead piles. For piling, fumigants are applied near the top which is then covered with a cap of coal-tar cement and fiberglass mesh. Fumigants have also been effectively used in horizontal members, particularly Douglas-fir. As a result, the use of fumigants to treat Douglas-fir poles, piles, and solid or laminated timbers with internal decay is common technology. However, fumigation treatments are not as long-lasting in Southern Pine poles and timbers as in Douglas-fir. The fumigants sodium N-methyl dithiocarbamate (Vapam), methylisothiocyanate (MITC-Fume), and trichloronitromethane (chloropicrin) are currently registered with the EPA for application to wood. They are normally applied by specialized personnel and only in exterior wood members. Re-treating of wood structures after 9 to 10 years is necessary because fumigants dissipate.

The control of insects by wood fumigants is uncertain. Vapam may control subterranean termites, but it has been noted that carpenter ants can reinfest wood shortly after fumigant application, perhaps because of the low concentration of the fumigant.

Piling

Piling again presents a special situation. Because piling is usually cut-off after driving, top decay results. This cut-off area should be treated (flooded) with hot creosote (150°F to 200°F), pentachlorophenol in diesel oil, or copper naphthenate, none of which penetrates the wood deeply. A water-shedding cap is then applied.

Materials used for capping piles include coal tar-roofing cement, galvanized metal, heavy roofing felt, heavy plastic, Noah's pitch, hot asphalt, and preservative-treated plywood. One effective capping device is coal tar-roofing cement held in place by a fiberglass mesh cloth. To cap a pile by this method, trowel a thick layer (½ in.) of cement on top, place two layers of fiberglass mesh on the cement, nail the mesh to the pile, and finish with an additional coat of cement. This patch remains flexible and resists water penetration into the untreated wood below.

Galvanized metal, roofing felts, and plastic sheets make effective caps when applied with chemical treatments. Without a preservative, condensation or leaks can create ideal conditions for decay beneath the cap. The material should be cut with at least a 2-in. overlap to permit the edges to be folded down and fastened to the pile sides with galvanized roofing nails or bands.

Preservative-treated plywood makes a simple and effective capping device. Two narrow strips of treated wood are nailed to the pile top, and the plywood cap, cut to a slightly larger diameter than the pile, is attached to the strips. The strips permit air to circulate beneath the cap, thus keeping the wood dry.

Because pile caps on working piers are often damaged or pulled off by hawsers, it is helpful to treat the pile top with a water-soluble preservative, such as boron, before cap installation. These treatments remain inactive as long as the cap remains effective, but they are activated whenever water penetrates.

Chapter 9
Specific Applications for Treated Wood

Specific Applications for Treated Wood

Numerous applications are available for treated wood products. Some are discussed here. Because many of the basic principles and concepts of deterioration and preservatives are discussed in previous chapters, the reader is encouraged to consult those sections as well.

Decks

Decks are discussed in several locations throughout this publication. Another publication gives additional detailed information.[18] Several important aspects should be considered when constructing wood decks. It is important to first select material of the grade that will provide the aesthetic characteristics needed. Low grades of lumber will have numerous knots and irregular edges where the bark has been removed.

Decks present a particularly severe exposure for wood and finishes and therefore require special consideration. Most wood members to be finished are in a horizontal or flat position. These horizontal surfaces are often exposed to the direct rays of the sun and tend to collect moisture, so the weathering process is greatly accelerated. As wetting and drying occurs, swelling and shrinking leads to checks that tend to enlarge rapidly into cracks and, along with the end-grain surfaces, tend to retain moisture. The conditions for decay and insect attack because of the presence of moisture are thereby greatly improved. Any film-forming finish (paint, solid-color stain, varnish) is subjected to excessive stress because of the continuous shrinking and swelling of the wood that results from changes in the moisture content. Furthermore, the finish is subjected to abrasive wear, particularly in high-traffic areas. For these reasons, film-forming finishes should never be used on decks.

As a result of the severe weather conditions that decks are exposed to, lumber that has been pressure treated with the waterborne preservatives or naturally durable wood such as redwood or western redcedar is generally used for the above ground portion. All posts or other members in soil contact should be pressure treated with the appropriate preservative at retention levels recommended for soil contact. In purchasing preservative-treated wood products, be sure to look for the quality mark or tag as discussed in Chapter 7.

Deck materials are left to weather naturally or finished in a number of ways. Some waterborne-preservative-treated wood already contains a water-repellent preservative and a coloring material, usually brown. Water-repellent preservatives help reduce checking and cracking of the wood.

Because most lumber treated with waterborne preservatives is shipped wet, about 3 to 5 percent shrinkage occurs as it dries in above ground applications. Spacings between deck boards or other joints must be made accordingly. For example a ¼-in. space between two 6-in. deck boards will widen to about ½ in. as the lumber dries. On the other hand, if the lumber is thoroughly dried before installation, it will swell in wet weather. Buckling may result. As lumber dries, it also tends to cup, bow, and twist if left unrestrained. Therefore, the material should be securely fastened in place with spiral or ring-shanked nails. Decking screws may also be used. When moisture is present and ferrous fasteners are used in conjunction with waterborne-preservative-treated wood, corrosion will occur. For decks, fasteners well treated by the hot-dip galvanizing process are suggested. Electroplated nails will corrode and should not be used. For additional protection and prevention from rust stains, stainless steel nails and screws are available and preferred.

Because of the severe conditions for fully exposed decks, a water-repellent preservative or a semi-transparent penetrating oil-based stain may provide the best finishing solution even on wood that has been pressure-treated with preservatives. Special formulations specifically made for decks are available. These penetrating finishes should be applied as soon as the wood surface is dry. Deck finishes, at least those that are water-repellent preservatives, may be somewhat shorter lived than paint, but they are more easily renewed. These penetrating finishes should be renewed annually; spring is usually the best time. Light-colored penetrating stains will last longer than dark stains on flat surfaces subject to traffic because light colors show the least contrast in grain color as wear occurs. Decks will need refinishing every 1 to 2 years. To refinish, a simple cleaning with a bristle brush is usually adequate before application of the water-repellent or penetrating finish. Paint and solid-color stains, particularly in these applications, are likely to peel and should not be used. Laborious scraping and sanding before refinishing will usually be required. Thus, paint, varnish, and solid-color stains are not good finishes for fully exposed decks. If they are used, the wood must be dry all the way through before application. A more thorough discussion on deck design, construction, and maintenance is available.[18]

Picnic Tables and Outdoor Furniture

Picnic tables and outdoor furniture are commonly constructed from wood treated with waterborne preservatives or naturally durable wood. Recommendations for selecting material and fastening are the same as given for decks. Be sure to read the appropriate Consumer Information Sheets given in Chapter 11 where some concern is expressed if food will be in contact with a treated-wood surface. It is recommended that wood treated with the waterborne preservatives CCA or ACZA not be used for cutting boards or countertops. In these applications, food is in intimate contact with the surface. Picnic tables are primarily used for serving prepared food. However, if you wish to minimize contact with treated wood, naturally durable material could be used for the tops and seats and treated wood for the legs and cleats. If treated wood has been used for the tops and seats, a sealant of 20 percent varnish in mineral spirits will reduce exposure to the preservative. Additional information concerning the suitability of CCA is presented in Chapter 11.

Landscape Timbers

Landscape timbers can take various forms. One common form is a piece about 3-½ in. wide with two flat faces and two round faces. This product is generally intended for light residential uses such as raised flower beds, borders, and low retaining walls. These pieces are generally produced from small diameter trees or the core left when a larger log is peeled for plywood. If the pieces contain a large percentage of heartwood, they are probably not well treated. There are no AWPA standards for treatment of landscape timbers and many treaters expressly disclaim adequacy of treatment. Thus, if long-term performance is important or in critical structural applications, select material specifically treated to AWPA Standard C2 (Lumber and Timbers).

Heavy timbers and some specialty treated products may also be used for landscape purposes. When using these larger members for critical applications such as major retaining walls, a procurement and inspection system as discussed in Chapter 7 should be followed.

Gardening Applications

Lumber treated with waterborne preservatives is used for a number of gardening and agricultural applications. Some uses include grape or tomato stakes, vineyard supports, orchard props, mushroom trays, trellises, raised vegetable beds, retaining walls, and compost beds. Unlike the oil-based pentachlorophenol and creosote, the material is virtually nontoxic to plants. Problems have not been reported, and a recent study shows that arsenic levels in soil samples taken near CCA-treated wood are within the normal range for all soils.[11] Chapter 11 presents additional information.

Playgrounds

Wood is commonly used in the construction of playground equipment. When exposed to the weather, this equipment needs to be protected from decay and insect attack. Naturally durable woods, as well as materials treated with waterborne preservatives, are commonly used for playground equipment.

The current AWPA Standard C17 says playground equipment treated with CCA must be free of visible surface deposits. In addition, AWPA Standard C1 says all material treated with waterborne preservative shall be supplied free of visible surface deposits. California has ordered that State funds cannot be used to purchase new wooden playground equipment treated with arsenic, penta, or creosote unless the equipment is free of visible surface residues. The law also prevents State funds from being spent on maintenance of wooden equipment treated with these preservatives unless a sealant (see section on Picnic Tables) is applied to prevent children from contacting the chemical. At least one material has been developed and accepted by that state to encapsulate the preservative-treated wood.[21]

Marine Framing and Seawalls

Treated lumber and timbers are often used for marine framing and seawalls. Both creosoted and CCA-treated material or a combination of the two can be used. In fresh water applications, material treated with 0.4 lb/ft^3 of CCA or that which is intended for soil contact can be used. In salt water applications, a retention of 2.50 lb/ft^3 of CCA is required.

Lumber and timber of those pine species that have no heartwood penetration requirement and are treated with waterborne preservative should be free of exposed heartwood on the sawn faces. Such material is available commercially under the designations Marine Framing or Seawall Grade. Cutting of material at the job site should be minimized because of possible exposure of poorly treated wood to marine borers. When cutting or other machining is necessary, exposed surfaces should be flooded with a preservative such as copper naphthenate.

Permanent Wood Foundations

Southern Pine lumber and plywood specially treated with CCA and Douglas-fir treated with ACZA can be used to build full basements, crawl space foundations, or an underfloor plenum system. Permanent Wood Foundation (PWF) is the correct industry terminology. These foundations can be installed during freezing weather, thus making winter construction feasible. Permanent Wood Foundation is accepted by the major model building codes, Federal agencies, and lending, home warranty, and fire insurance institutions.

Because permanent wood foundations are a critical application for wood, a special AWPA Standard C22 was developed. The wood is treated with 0.6 lb/ft^3 of preservative compared with 0.4 lb/ft^3 for standard lumber used in soil contact. The FDN in the quality mark is an abbreviation for foundation, indicating that the material is suitable for that application.

For any PWF material applied below grade, stainless steel fasteners should be used. Hot-dipped galvanized fasteners may be used above grade, providing excessive moisture is not present. Standard ferrous fasteners will corrode.

Shakes or Shingles

Wooden shingles or shakes have typically been made of naturally durable species such as Western redcedar. Sawn shakes produced from southern pines pressure treated with CCA are now available. The industry is producing two different grades, and an inspection program is in place to certify that the shingles are properly treated. A proposed standard specifically written for this product is under consideration by AWPA.

Log Structures

Log structures present a challenge in wood preservation. They are particularly susceptible to decay because of the deep seasoning checks that occur on the surface of large wood members. These checks allow moisture to penetrate into the wood member, and decay results. Excessive end grain is also exposed, particularly on the corners of log structures. This end grain, as well as the notching of the logs, allows for easy penetration of moisture and subsequent high decay potential.

Old structures that are still sound were usually constructed with attention to moisture-proofing details. Also, naturally durable woods such as white oak, walnut, heartwood of white pine, and cedar were usually used, particularly for the lower courses of logs where decay is most prevalent. However, even these woods will succumb to decay with prolonged exposure to moisture.

Many log structures available today are made with small diameter logs containing large percentages of sapwood. Sapwood is susceptible to decay and insect attack. As a result, proper construction is essential in preventing deterioration. Good practices include (1) adequate roof overhang with properly hung and maintained gutters and downspouts; (2) wrap around porches that protect the lower portion of exterior walls; (3) avoidance of tall exposed walls of logs; (4) roof or other appropriate vents, if attic space is present; (5) good drainage and ventilation around the foundation; (6) proper venting of showers, baths, and dryers; (7) adequate clearance between the soil and lower logs; and (8) most important of all, proper design so that any moisture contacting the log surface will run down and off the logs rather than becoming trapped. Trapped moisture is certain to result in wood decay.

These practices, if properly done, will usually limit the amount of deterioration occurring. Remember, that the log usually forms both the inside and outside of any walls. Therefore, preservatives applied by pressure or dipping prior to construction will be on both the inside and outside of the structure.

For existing structures, even those that are well weathered, the application of a water-repellent preservative to the exterior can help prevent further deterioration. Any decayed wood should be removed, and the newly exposed wood liberally treated after the moisture source has been eliminated and the wood dried. To facilitate treatment of hard-to-reach areas that may have some decay, ¼-in. holes can be drilled in the wood and filled, preferably several times, with preservative solution that will diffuse into the adjacent wood. The holes should then be plugged with preservative-treated wood dowels.

Note: Most preservatives should not be used indoors. Always check manufacturer's recommendations and Material Safety Data Sheets for proper use of all preservatives.

The use of borates to control wood decay is a relatively new development in the United States and may be particularly useful on log structures with moisture problems. The material is manufactured for brush or spray application or as a rod that can be inserted into holes bored in the wood. The material diffuses through wet wood and provides protection against decay and wood-destroying insects. This material is available through distributors and can be applied by professionals, such as pest control operators, log cabin manufacturers, and pole treaters.

Railroad Ties

Used railroad ties are commonly available for landscape and other construction purposes. These ties were initially treated with creosote to railroad specifications and have been removed from service because of a railroad line being closed, excessive mechanical wear at the tie plate and spike loosening, excess splintering or checking, and decay. Obviously, the quality can vary substantially. The sound treated wood remaining is likely to be very durable. Care should be exercised in making certain that the used ties are acceptable for the intended end-use. Remember, creosote has a pronounced odor, cannot be painted, and should not be used when skin contact is likely.

If new railroad ties or equivalent-sized timbers are being purchased ask how the material was treated and if a warranty is available. Penetration of the preservative into large timbers or ties is sometimes difficult. It can be checked by boring or cutting in the middle of the piece. If the material is being offered "as is" and carries no quality mark or tag identifying the treater, it is probably best to avoid this material.

Disposal of wood materials in many landfills is restricted; therefore, the accumulation of excess quantities of treated wood should be avoided. The Consumer Information Sheets (Chapter 11) strictly prohibit the burning of treated wood except in approved incinerators.

Chapter 10
Finishing Treated Wood

Finishing Treated Wood

Wood products appropriately treated with preservatives for a specified end-use and properly installed will withstand decay, insect, and marine borer attack for extended periods. In addition to these biological organisms, wood in exposed conditions is subject to weathering. Weathering occurs as wood is exposed to the direct rays of the sun and alternating wetting and drying cycles caused by precipitation. With these forces, small checks tend to enlarge into cracks on both flat surfaces and end grain. Moisture accumulates rapidly in these openings and on end grain, and the process continues to accelerate. If the cracks become deep enough to penetrate the preservative shell, decay and insect attack can occur on the interior, untreated portion of the wood products. The wood also turns a gray color or it can mildew and turn dark.

This weathering phenomena is particularly troublesome on wood products treated with waterborne preservatives such as CCA and ACZA or untreated naturally durable wood. Virtually all treated lumber, plywood, posts, small poles, and timbers available at retail lumber yards are treated with waterborne preservatives and subject to weathering if exposed. Some mildew or mold organisms can develop on the surface of waterborne-preservative-treated lumber. Therefore, if aesthetics are important and weathered wood is to be avoided, the application of wood finishes should be considered.

Wood Characteristics Affecting Finishes

Grain characteristics of the particular species being used is another factor that must be considered when selecting a finish. Much of the preservative-treated lumber available in the eastern half of the United States is Southern Pine. Lumber from Southern Pine as well as Douglas-fir, which is more common in the West, is dominated by a flat-grain pattern and wide latewood bands. These two characteristics tend to cause paint to peel. As a result, these two lumber groups have the least desirable paint-holding characteristics. Hemlock and a group called hem-fir and ponderosa pine in the West and red and Jack pine in the North have somewhat better paint-holding characteristics.

Forest Products Laboratory research has shown that clear water repellents, semitransparent stains, and paints perform much better on CCA-treated wood than they do on untreated lumber from the same species. The chromium in the CCA reduces the degrading effect of weather.

Waterborne Preservatives

Because of the conditions to which most waterborne-preservative-treated wood is exposed and the grain characteristics of the more common species, special precautions in the selection, application, and maintenance of a wood finish must be considered. For fully exposed preservative-treated wood or naturally durable material, a water-repellent preservative or a semitransparent oil-based penetrating stain containing a mildewcide may provide the best finishing solution. Some preservative-treated wood is being produced with a water repellent and even a coloring material, usually brown, added at the time of treatment. Be sure to check with the supplier concerning the details. Film-forming finishes, such as paint, may be used in protected areas, but are not recommended for fully exposed areas.

Water-Repellent Preservatives

A water-repellent preservative acts as a natural finish. The treatment reduces warping and checking, prevents water staining, and helps control mildew growth. Water-repellent preservatives contain a fungicide, a small amount of wax as a water repellent, a resin or drying oil, and a solvent such as turpentine, mineral spirits, or paraffinic oil. Those used for decks often contain an ultraviolet light stabilizer or absorber. This stabilizer prevents the sun from slowly degrading the surface of wood, releasing fibers and groups of fibers. The wax reduces the absorption of liquid water by the wood (**Fig. 10-1**), and the preservative prevents wood from darkening (graying) by inhibiting the growth of mold and mildew. The finish will vary in color, depending upon the wood color, but will usually weather to a clean, golden tan. For severe exposures, the finish should be renewed annually; spring is usually the best time. Some waterborne formulations are available.

Semitransparent Penetrating Stains

Semitransparent penetrating oil-based stains are pigmented water repellents or water-repellent preservatives. They penetrate the wood surface to a degree, are porous, and do not form a surface film like paint. Thus, they do not totally hide the wood grain. As a result, the stains will not blister or peel even if moisture penetrates the wood.

Penetrating stains are oil-based (or alkyd-based), and those used on decks should contain a fungicide (preservative or mildewcide), a water repellent, and often an ultraviolet light stabilizer. Make certain that the manufacturer's label indicates that the finish will resist mildew or that the product contains a specific mildewcide. All fungicides are not effective against mildew. Light-colored penetrating stains will also last longer than dark ones on flat surfaces subject to traffic because light stains show the least contrast in grain color as wear occurs. On the other hand, dark colors can reduce or eliminate the green color of some waterborne-preservative-treated wood from showing through and even-out the substrate appearance. Semitransparent stains may last up to 4 years for deck applications.

Latex-based (waterborne) stains are also available, but they do not penetrate the wood surface as well as do their oil-based counterparts. Newer latex formulations that may provide better penetrating characteristics are being developed.

To refinish a water-repellent preservative or semitransparent stain, a simple cleaning of the wood with a bristle brush is usually adequate before applying the new finish. In applying a water-repellent preservative or semitransparent stain, the surface should be clean, dry, and free of mildew. The coatings may be applied by brush, spray, roller, or pad. However, brushing the finish generally gives the best penetration and performance because it tends to work the finish into surface checks or other small openings that have developed. If spraying is used to apply a pigmented finish, back brushing the wet coating with a brush or push broom will even out the finish and eliminate drip and lap marks. Be certain to apply the finish at the recommended coverage rate. Finishes specially formulated for decks are designed to penetrate the wood surface. Excess buildup of the material will form a film that can ultimately peel or crack. Over-applied stains can result in sticky surfaces as a result of improper finish drying, and water repellents can become overly waxy, sticky, or slick.

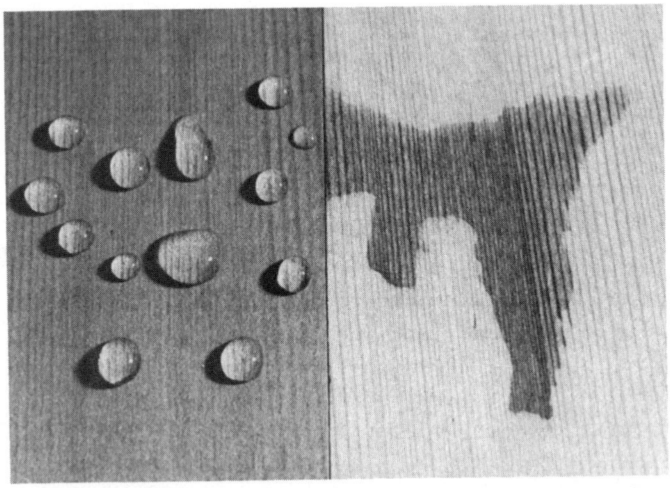

Figure 10-1. *Wood surface brush-treated with water repellent (left). The treated surface resisted penetration by liquid water, whereas the untreated wood surface (right) absorbed water quickly.*

Coatings should also be applied under favorable weather conditions following manufacturer's recommendations. Solvent-borne coatings should be applied when the temperature is at least 40°F or greater. For water-based products, the temperature should be at 50°F or greater for at least 24 hours after application. Ideally, the coatings should have 12 to 24 hours to cure before any precipitation occurs.

Paints

Paints, varnishes, and solid-color stains, applied in exposed applications, such as decks, are likely to peel. These finishes should not be used on exposed decks. Laborious scraping and sanding before refinishing will usually be required to restore such finishes. However, paints may be successfully used on roof-protected porch floors and vertical surfaces such as siding, trim, plywood foundations, and lattice work. The best procedure is to treat the wood with a paintable water-repellent preservative first. After the treatment dries, a primer and two topcoats should be applied. Porch enamel is especially formulated to resist abrasion and wear.

Because of their low resin content, solid-color stains should never be used on flat surfaces such as decks and porches. Varnish will also fail or peel in a very short time and should not be used. Decks represent a combination of the most adverse conditions, refer to Chapter 9 for finishing and other details.

Note: Wood that is pressure treated with waterborne preservatives often contains large quantities of water when shipped to retail lumber yards. Therefore, care should be exercised to make certain that the lumber is dry before finishing. For penetrating stains, at least the surface should be thoroughly dry. If paint is to be applied, the piece should be dry throughout. Drying may take days or weeks, depending on how wet the wood is initially and drying conditions. Air drying in place is acceptable, although some shrinking, warp-

*ing, and checking may result (**Fig. 5-1**). Finishes applied to wet wood will fail prematurely.*

Refinishing

On well-weathered decks, the original bright color of the wood can be restored by application of commercial products called deck cleaners, brighteners, or restorers. These products may remove the weathered wood surface, and care should be exercised not to remove excess wood. Color can also be restored using a liquid household bleach containing 5 percent sodium hypochlorite. The bleach is usually diluted with water (1 part bleach, 3 parts water) before it is applied to the deck. The bleach solution should be rinsed from the deck with water. If the deck is to be finished after cleaning, allow 1 to 2 days drying time.

In some sections of the country, contractors or rental agencies with power washing equipment may be available to help restore weathered wood ranging from decks and fences to wood-shingled roofs. Power washing can be damaging to the wood and must be done with care.

Oil-Based Preservatives

When creosote or pentachlorophenol in a heavy oil solvent with low volatility is used to pressure treat wood, successful finishing is impossible. The surface is generally oily and dark, and the dark color will usually bleed through any paint. Poor adhesion is also likely for all film-forming finishes. Because of the oil base, wood treated with these preservatives does not tend to weather as rapidly as that treated with the waterborne preservatives. However, if the wood has weathered for several years, it can sometimes be stained, but rarely painted. It is best to finish a small area first and expose it to direct sunlight on warm to hot days to determine the performance of the finish.

Wood pressure treated with a water-repellent preservative is loaded with relatively large quantities of the preservative and solvent, and the solvent must be completely removed before painting is attempted. Solvents vary in their volatility, and evaporation may take from 1 to 2 years. Even with special drying schedules, the complete paintability of the wood may not be restored. However, this process should not be confused with the brush, spray, or dip treatment of a water-repellent preservative, which serves to enhance paint durability and wood performance.

For a detailed discussion on wood finishing, the reader is encouraged to consult Exterior Wood Finishes in the South: Selection, Applications, and Finishes[7] or Finishing Wood Exteriors: Selection, Application and Maintenance.[6]

Chapter 11
Safety and Environmental Issues

Page

Safety and Environmental Issues

Wood preservatives are considered pesticides and their use is closely regulated by the EPA. Some preservatives are available to anyone as over-the-counter purchases, and others are considered restricted-use pesticides and available only to certified applicators. Certified applicators are individuals who have received specialized training on the proper use of pesticides. Some preservatives may require special handling or disposal under the Resource Conservation and Recovery Act (RCRA). Regulations vary by region.

It is important to recognize that the wood preservatives (pentachlorophenol, creosote, CCA, and ACZA) most commonly applied by pressure treatment are all restricted-use pesticides, but wood treated with these chemicals is not restricted. To provide for the maximum personal safety and minimize potential environmental damage, wood preservatives should be applied as specified on the label. Wood treated with CCA, ACZA, pentachlorophenol, or creosote should be used as directed in the Consumer Information Sheets as developed by EPA. These sheets should be available from the supplier of the materials.

After CCA is impregnated into wood, its chemical composition changes and becomes leach resistant. This chemical change (often called a fixation process) occurs as the wood dries and is accelerated by heat. Wood freshly treated with CCA that has not had time to dry or has not been exposed to warm temperatures will contain uncured CCA. Particular caution should be paid when handling wet CCA-treated material purchased during the winter months in temperate climates. Cautions include wearing rubber gloves when handling the material, avoiding machining, and washing afterwards. Similar cautions should be used when handling ACZA-treated wood.

Another prudent practice is to use preservatives or preservative-treated wood products only when the end-use requires use. Using preservatives in other applications adds additional pesticide to the environment and increases costs.

Finally, alternate preservatives such as borates, copper naphthenate, and ACQ are available for some applications. Currently these preservatives are not restricted-use pesticides and are reportedly effective if used as directed.

Health-Related Issues

Wood treated with CCA has been used successfully for many years because the surface is not oily and the wood is durable. Treated wood is often used where people, children, or animals can come in contact with it. However, the basic chemical components of chromium and arsenic have caused many health-related and environmental questions, especially when the material is used in applications such as picnic tables and lawn furniture, gardening and agricultural uses, and playground equipment. The issue concerns material that can leach from the wood and contaminate food, feed, and soil and if it comes into dermal or oral contact, especially by children, chemicals may be absorbed. As a result, several studies have been published and some are reviewed here.

Concerning leachability of the material, Arsenault indicated that

> Treated wood poles in service in various soil types from 26 to 32 years did not show significant loss of any pesticide component due to leaching or other factors. Furthermore, soil samples collected at various distances from poles treated with CCA of a type most comparable to current CCA showed an average (four poles) arsenic concentration of 53.0 ppm at the pole and decreasing to 16.3 ppm approximately one foot from the pole. The background soil arsenic level for the area was 14.2 ppm. Arsenic is present in soils naturally with a range of levels from about 1.0 ppm to about 8000 ppm and an average of 5-6 ppm. CCA in current use has been improved to provide even greater performance than that reported herein.[2]

Although permitted under EPA guidelines, some controversy exists about the use of CCA-treated wood in raised-bed gardening applications, regarding leaching of small amounts of chromium and arsenic into soil and possible uptake by plants. One study on the use of CCA-treated grapestakes showed no uptake.[16]

A recent but limited study of soil samples from raised bed gardens constructed with CCA-treated landscape timbers also suggests that this is a defendable practice. Analysis of soil samples at several distances from the treated wood led Finch and Dainello to conclude that leaching of CCA was not a problem in this application.[11]

Other rigorous scientific studies have been proposed that should give definitive answers to this issue.

In another study concerning the use of CCA-treated wood for playground equipment, it was reported that

> No adverse effect on health of users has been reported. On theoretical grounds the health hazard, if any, should be only small because commercial CCA products in treated pine undergo fixation within a few days and the components become only sparingly soluble from wood in contact with water. However, the fact that the pine is treated with copper, chromium and arsenic compounds has raised some doubts as to its safety for use in children's playgrounds. The surface of the unweathered material may carry traces of soluble preservative or deposits of reacted components.[15]

In this investigation, commercially treated rounds were examined, and the amount of preservative on the surfaces was assessed.

> The health hazard appears to be small. However, because a considerable variation is possible in the extent of fixation of the copper, chromium, and arsenic components and the uncertainty as to the toxicity of the fixed arsenic to humans, a cautious approach is warranted. Thorough washing by scrubbing of the treated surface, particularly the intact ends, are recommended when the treated material is accessible to children.[15]

Research has shown that wood treated with CCA should present a minimal health risk when properly handled and used. However, for individuals concerned about the presence of chromium and arsenic, some additional actions or alternatives exist. First, thoroughly scrubbing and washing the surface, and especially the ends of treated lumber, will help remove any surface residues. It has also been suggested that the surface can be sealed with a 20- to 30-percent varnish mixture in mineral spirits. If exposed to the weather, this treatment would probably require renewal on a yearly basis. For treated wood in raised-bed gardens, plastic film has been used to isolate the wood preservative from growing plants. Naturally durable woods can also be used in place of treated wood, or alternative preservatives that do not contain chromium or arsenic can be considered for some applications.

Disposal

Guidelines for disposal of treated wood are provided in the Consumer Information Sheets that are available from suppliers of treated products. Important points are repeated here:

- Treated wood should be disposed through ordinary trash collection or burial.
- Treated wood should not be burned in open fires, stoves, fireplaces, or residential furnaces, or boilers.

With rapidly-filling landfills and growing sentiment against disposal of any biomass in landfills, disposal of spent treated wood may become more difficult in the future. The use of spent treated wood as biomass fuel is one option that is being developed. The EPA Consumer Information Sheets indicate that "Treated wood from commercial or industrial use (e.g., construction sites) may be burned only in commercial or industrial incinerators or boilers in accordance with State and Federal regulations." Materials treated with creosote, such as railroad ties, or pentachlorophenol-treated wood are being successfully disposed in this manner. Research on the incineration, bioremediation, physical, and chemical treatment of used preservative-treated products is on-going and yielding important results. At this time, it is probably prudent to consider and possibly develop a life-cycle plan that includes the proper management of treated wood during its primary use followed by one or more secondary uses and disposal options. For example, material treated and used for several years as small poles or railroad ties may find a secondary use as posts or landscaping purposes.

Preservative Poisoning and First Aid

Certain nonrestricted preservatives are available to general consumers. The user of these preservatives is bound by law to follow label directions. Following label directions is essential to safe, effective, and environmentally sound pesticide application. First aid procedures are generally given on the label. Be sure to read the label before applying the preservative.

In the event of an acute preservative or other pesticide poisoning, certain emergency procedures are recommended.[25] Step one is to prevent further exposure and make sure the victim is breathing. Then call emergency medical personnel. While help is on the way, maintenance of vital signs is imperative, and cardiopulmonary resuscitation techniques may be required. Be ready to provide medical personnel with a copy of the pesticide label.

For personal safety, be sure you know and have a phone number for the closest poison information center. Also, be sure to contact your state pesticide program for local and detailed information on safety. State pesticide programs are usually administered by the Cooperative Extension Service, a part of the Agricultural Complex in each land grant institution.

Consumer Information Sheets

The EPA-approved information sheets for each of the three major groups of wood preservatives are being provided by producers to consumers of treated-wood products. These sheets, reproduced here, provide users with information about the preservative and the use and disposal of treated-wood products. As a minimum, anyone involved with preservative-treated wood products should become familiar with these information sheets and follow the advice therein.

Inorganic Arsenical Pressure-Treated Wood

CONSUMER INFORMATION

This wood has been preserved by pressure-treatment with an EPA-registered pesticide containing inorganic arsenic to protect it from insect attack and decay. Wood treated with inorganic arsenic should be used only where such protection is important.

Inorganic arsenic penetrates deeply into and remains in the pressure-treated wood for a long time. Exposure to inorganic arsenic may present certain hazards. Therefore, the following precautions should be taken both when handling the treated wood and in determining where to use or dispose of the treated wood.

USE SITE PRECAUTIONS

Wood pressure-treated with waterborne arsenical preservatives may be used inside residences as long as all sawdust and construction debris are cleaned up and disposed of after construction.

Do not use treated wood under circumstances where the preservative may become a component of food or animal feed. Examples of such sites would be structures or containers for storing silage or food.

Do not use treated wood for cutting-boards or countertops.

Only treated wood that is visibly clean and free of surface residue should be used for patios, decks, and walkways.

Do not use treated wood for construction of those portions of beehives which may come into contact with the honey.

Treated wood should not be used where it may come into direct or indirect contact with public drinking water, except for uses involving incidental contact such as docks and bridges.

HANDLING PRECAUTIONS

Dispose of treated wood by ordinary trash collection or burial. Treated wood should not be burned in open fires or in stoves, fireplaces, or residential boilers because toxic chemicals may be produced as part of the smoke and ashes. Treated wood from commercial or industrial use (e.g., construction sites) may be burned only in commercial or industrial incinerators or boilers in accordance with state and Federal regulations.

Avoid frequent or prolonged inhalation of sawdust from treated wood. When sawing and machining treated wood, wear a dust mask. Whenever possible, these operations should be performed outdoors to avoid indoor accumulations of airborne sawdust from treated wood.

When power-sawing and machining, wear goggles to protect eyes from flying particles.

After working with the wood, and before eating, drinking, and use of tobacco products, wash exposed areas thoroughly.

If preservatives or sawdust accumulate on clothes, launder before reuse. Wash work clothes separately from other household clothing.

Pentachlorophenol Pressure-Treated Wood

CONSUMER INFORMATION

This wood has been preserved by pressure-treatment with an EPA-registered pesticide containing pentachlorophenol to protect it from insect attack and decay. Wood treated with pentachlorophenol should be used only where such protection is important.

Pentachlorophenol penetrates deeply into and remains in the pressure-treated wood for a long time. Exposure to pentachlorophenol may present certain hazards. Therefore, the following precautions should be taken both when handling the treated wood and in determining where to use and dispose of the treated wood.

USE SITE PRECAUTIONS

Logs treated with pentachlorophenol should not be used for log homes. Wood treated with pentachlorophenol should not be used where it will be in frequent or prolonged contact with bare skin (for example, chairs and other outdoor furniture), unless an effective sealer has been applied.

Pentachlorophenol-treated wood should not be used in residential, industrial, or commercial interiors except for laminated beams or building components which are in ground contact and are subject to decay or insect infestation and where two coats of an appropriate sealer are applied. Sealers may be applied at the installation site. Urethane, shellac, latex epoxy enamel, and varnish are acceptable sealers for pentachlorophenol-treated wood.

Wood treated with pentachlorophenol should not be used in the interiors of farm buildings where there may be direct contact with domestic animals or livestock which may crib (bite) or lick the wood.

In interiors of farm buildings where domestic animals or livestock are unlikely to crib (bite) or lick the wood, pentachlorophenol-treated wood may be used for building components which are in ground contact and are subject to decay or insect infestation and where two coats of an appropriate sealer are applied. Sealers may be applied at the installation site.

Do not use pentachlorophenol-treated wood for farrowing or brooding facilities.

Do not use treated wood under circumstances where the preservative may become a component of food or animal feed. Examples of such sites would be structures or containers for storing silage or food.

Do not use treated wood for cutting-boards or countertops.

Only treated wood that is visibly clean and free of surface residue should be used for patios, decks, and walkways.

Do not use treated wood for construction of those portions of beehives which may come into contact with the honey.

Pentachlorophenol-treated wood should not be used where it may come into direct or indirect contact with public drinking water, except for uses involving incidental contact such as docks and bridges.

Do not use pentachlorophenol-treated wood where it may come into direct or indirect contact with drinking water for domestic animals or livestock, except for uses involving incidental contact such as docks and bridges.

HANDLING PRECAUTIONS

Dispose of treated wood by ordinary trash collection or burial. Treated wood should not be burned in open fires or in stoves, fireplaces, or residential boilers because toxic chemicals may be produced as part of the smoke and ashes. Treated wood from commercial or industrial use (e.g., construction sites) may be burned only in commercial or industrial incinerators or boilers rated at 20 million BTU/hour or greater heat input or its equivalent in accordance with state and Federal regulations.

Avoid frequent or prolonged inhalation of sawdust from treated wood. When sawing and machining treated wood, wear a dust mask. Whenever possible, these operations should be performed outdoors to avoid indoor accumulations of airborne sawdust from treated wood.

Avoid frequent or prolonged skin contact with pentachlorophenol-treated wood. When handling the treated wood, wear long-sleeved shirts and long pants and use gloves impervious to the chemicals (for example, gloves that are vinyl-coated).

When power-sawing and machining, wear goggles to protect eyes from flying particles.

After working with the wood, and before eating, drinking, and use of tobacco products, wash exposed areas thoroughly.

If oily preservatives or sawdust accumulate on clothes, launder before reuse. Wash work clothes separately from other household clothing.

Creosote Pressure-Treated Wood

CONSUMER INFORMATION

This wood has been preserved by pressure treatment with an EPA-registered pesticide containing creosote to protect it from insect attack and decay. Wood treated with creosote should be used only where such protection is important.

Creosote penetrates deeply into and remains in the pressure-treated wood for a long time. Exposure to creosote may present certain hazards. Therefore, the following precautions should be taken both when handling the treated wood and in determining where to use the treated wood.

USE SITE PRECAUTIONS

Wood treated with creosote should not be used where it will be in frequent or prolonged contact with bare skin (for example, chairs and other outdoor furniture) unless an effective sealer has been applied.

Creosote-treated wood should not be used in residential interiors. Creosote-treated wood in interiors of industrial buildings should be used only for industrial building components which are in ground contact and are subject to decay or insect infestation and wood-block flooring. For such uses, two coats of an appropriate sealer must be applied. Sealers may be applied at the installation site.

Wood treated with creosote should not be used in the interiors of farm buildings where there may be direct contact with domestic animals or livestock which may crib (bite) or lick the wood.

In interiors of farm buildings where domestic animals or livestock are unlikely to crib (bite) or lick the wood, creosote-treated wood may be used for building components which are in ground contact and are subject to decay or insect infestation if two coats of an effective sealer are applied. Sealers may be applied at the installation site. Coal tar pitch and coal tar pitch emulsion are effective sealers for creosote-treated wood-block flooring. Urethane, epoxy, and shellac are acceptable sealers for all creosote-treated wood.

Do not use creosote-treated wood for farrowing or brooding facilities.

Do not use treated wood under circumstances where the preservative may become a component of food or animal feed. Examples of such use would be structures or containers for storing silage or food.

Do not use treated wood for cutting-boards or countertops.

Only treated wood that is visibly clean and free of surface residues should be used for patios, decks, and walkways.

Do not use treated wood for construction of those portions of beehives which may come into contact with the honey.

Creosote-treated wood should not be used where it may come into direct or indirect contact with public drinking water, except for uses involving incidental contact such as docks and bridges.

Do not use creosote-treated wood where it may come into direct or indirect contact with drinking water for domestic animals or livestock, except for uses involving incidental contact such as docks and bridges.

HANDLING PRECAUTIONS

Dispose of treated wood by ordinary trash collection or burial. Treated wood should not be burned in open fires or in stoves, fireplaces, or residential boilers, because toxic chemicals may be produced as part of the smoke and ashes. Treated wood from commercial or industrial use (e.g., construction sites) may be burned only in commercial or industrial incinerators or boilers in accordance with state and Federal regulations.

Avoid frequent or prolonged inhalations of sawdust from treated wood. When sawing and machining treated wood, wear a dust mask. Whenever possible these operations should be performed outdoors to avoid indoor accumulations of airborne sawdust from treated wood.

Avoid frequent or prolonged skin contact with creosote-treated wood; when handling the treated wood, wear long-sleeved shirts and long pants and use gloves impervious to the chemicals (for example, gloves that are vinyl-coated).

When power-sawing and machining, wear goggles to protect eyes from flying particles.

After working with the wood and before eating, drinking, and use of tobacco products, wash exposed areas thoroughly.

If oily preservative or sawdust accumulate on clothes, launder before reuse. Wash work clothes separately from other household clothing.

Chapter 12
Sources of Information

Sources of Information

Many sources of information are available for individuals trying to learn about all types of construction wood products. These sources can be divided into two general categories: public-supported institutions and numerous industrial trade associations promoting the proper use of wood as a construction material.

Public Institutions

The USDA Forest Service, Forest Products Laboratory, Madison, Wisconsin, has information and technical expertise in wood processing and many publications that are available.

The land grant institutions, or equivalent, in each state are another major source of information. The department or college of forestry, or equivalent, is a logical starting place. More specifically, ask to speak to someone with wood technology responsibility. Individuals with responsibilities for buildings and construction within agricultural engineering may be equally valuable. A few institutions have a wood research laboratory. The type and breadth of these programs vary substantially by state.

State forestry commissions or departments of natural resources may also be of some help. Ask to speak to someone familiar with the wood manufacturing industry.

Trade Associations

Trade associations can be an especially valuable source of information to buyers, users, and specifiers of the products they represent. Several key associations follow.

The American Wood Preservers' Association (AWPA) maintains the standards for the industrial pressure-treating industry. Specific questions on preservative treatment processes and how products should be treated are best referred to this organization.

The American Wood Preservers Institute (AWPI) is primarily concerned with environmental issues and regulations. Printed materials are available.

The American Lumber Standard Committee (ALSC), although not strictly a Trade Association, is the organization that accredits agencies that oversee the labeling of wood products treated according to accepted AWPA standards.

The Southern Forest Products Association (SFPA) is a major trade association representing several aspects of the southern pines manufacturing industry. Of specific interest is lumber, treated products, and shakes. Numerous publications are available from the Southern Pine Marketing Council located at the same address.

The Western Wood Products Association (WWPA) is the West Coast and western species equivalent to the SFPA.

The Western Wood Preservers Institute (WWPI) represents companies specifically involved in treating western species. Several publications on western species are available.

The trade associations just mentioned and others can be reached at the following locations:

American Forest and Paper Association
1111 19 St., NW
Washington, DC 20036
Telephone 202/463-2762
Fax 202/463-2791

American Lumber Standard Committee
P.O. Box 210
Germantown, MD 20875-0210
Telephone 301/972-1700
Fax 301/540-8004

American Plywood Association
P.O. Box 11700
Tacoma, WA 98411
Telephone 206/565-6600
Fax 206/565-7265

American Wood Preservers' Association
P.O. Box 286
Woodstock, MD 21163-0286
Telephone 410/465-3169
Fax 410/465-3195

American Wood Preservers Institute
1945 Old Gallows Rd.
Suite 550
Vienna, VA 22182
Telephone 703/893-4005
Fax 703/893-8492

National Wood Window and Door Association
1400 East Touhy Avenue
Des Plaines, IL 60018
Telephone 708/299-5200
Fax 708/299-1286

Southern Forest Products Association
P.O. Box 52468
New Orleans, LA 70152
Telephone 504/443-4464
Fax 504/443-6612

Western Wood Preservers Institute
P.O. Box 2913
Vancouver, WA 98668
Telephone 206/693-9958
Fax 206/693-9967

Western Wood Products Association
Yeon Building
522 S.W. Fifth Avenue
Portland, OR 97204
Telephone 503/224-3930
Fax 503/224-3934

References

1 Agi, Jerry J. 1983. Structural evaluation of marine structures. *In*: Specialty conference on port modernization, upgrading and repairs. American Society of Chemical Engineers, New Orleans, LA; 1983 March 21-23: p. 662-677.

2 Arsenault, Robert D. 1975. CCA-treated wood foundations: A study of permanence, effectiveness, durability, and environmental considerations. *In*: Proceedings of the American Wood Preservers' Association. Vol. 71. p 126-147.

3 ASHRAE. 1989. Handbook, Fundamental I-P Addition. American Society of Heating, Refrigerating and Air Conditioning Engineers, Atlanta, GA.

4 AWPA. 1993. American Wood Preservers' Association Standards. American Wood Preservers' Association, Seattle, WA.

5 Campbell, R.N.; Clark, J.W. 1960. Decay resistance of baldcypress heartwood. Forest Products Journal. 20(5): 250-253.

6 Cassens, Daniel L.; Feist, William C. 1986. Finishing wood exteriors: Selection, application, and maintenance. Agric. Handb. No. 647. Washington, DC: U.S. Department of Agriculture. 56 p.

7 Cassens, Daniel L.; Feist, William C. 1991. Exterior wood finishes in the south: Selection, applications, and finishes. Gen. Tech. Rep. FPL-GTR-69. Madison, WI: U.S. Department of Agriculture, Forest Service, Forest Products Laboratory. 56 p.

8 Clark, Joe W.; Scheffer, Theodore C. 1983. Natural decay resistance of the heartwood of coast redwood *Sequoia sempervirens* (D. Don). Forest Products Journal. 33(5): 15-20.

9 CORRIM. 1976. Wood for structural and architectural purposes. Committee on Renewable Resources for Industrial Materials-National Research Council (CORRIM) Wood and Fiber, Madison, WI.

10 Davidson, H.L. 1977. Comparison of wood preservatives in Mississippi post study. Res. Note FPL-RN-01. Madison, WI: U.S. Department of Agriculture, Forest Service, Forest Products Laboratory. 16 p. [from unpublished 1990 update]

11 Finch, Calvin; Dainello, Frank. 1992. Arsenic leaching from CCA treated landscape timbers. Texas Agricultural Extension Service, San Antonio, TX.

12 Forest Products Laboratory. 1987. Wood handbook: Wood as an engineering material. Agric. Handb. 72. Washington, DC: U.S. Department of Agriculture; rev. 466 p.

13 Grove, Scott L. 1987. Copper naphthenate: An alternative wood preservative. Presented at the 41st annual meeting of the Forest Products Research Society; 1987 June 23; Louisville, KY. 14 p.

14 Hunt, George M.; Garrett, George A. 1967. Wood preservation. 3d ed. New York: McGraw Hill Book Company. 433 p.

15 Johanson, R.; Dale, F.A. 1973. Arsenic on the surface of round pine treated with Cu-Cr-as preservative. Holzforschung. 27(6): 187-189.

16 Levi, M.P.; Huisingh, D.; Nesbitt, W.B. 1974. Uptake by grape plants of preservatives from pressure treated posts not detected. Forest Products Journal. 24: 97-98.

17 MacLean, J.D. 1935. Manual of preservative treatment of wood by pressure. Misc. publication 224. Washington, DC: U.S. Department of Agriculture.

18 McDonald, K.A.; Falk, R.; Williams, R.S.; Winandy, J.E. [in press]. Wood decks: Materials, construction, and finishing. Forest Products Society, Madison, WI. __ p.

19 NFPA. 1991. National design specification for wood construction and supplement. National Forest Products Association, Washington, DC. 125 p.

20 NWWDA. [n.d.] Industry standard for water-repellant preservative treatments for millwork 1.5.4-93. National Woodwork Manufacturers Association, (currently National Wood Window and Door Association, DesPlaines, IL). 6 p.

21 Paint & Coatings Industry. 1991. Health officials voice concerns over playground equipment treated with wood preservatives. Paint & Coatings Industry. p. 28-29.

22 Powell, Douglas S. and others. 1992. Forest resources of the United States, 1992. Gen. Tech. Rep. RM-GTR-234. Fort Collins, CO: U.S. Department of Agriculture, Forest Service, Rocky Mountain Forest and Range Experiment Station. 132 p.

23 Scheffer, Theodore C. 1957. Decay resistance of western redcedar. Journal of Forestry. 55(6): 434-442.

24 SFPA. 1990. Pressure treated southern pine. Southern Forest Products Association, Kenner, LA. 14 p.

25 Whitford, Fred and others. 1992. Pesticides and personal safety. Purdue University Cooperative Extension Service, Purdue University, West Lafayette, IN. 15 p.

26 WWPI. 1993. Adapted from guide to the characteristics, use and specifications of pressure treated wood. Western Wood Preservers Institute, Vancouver, WA. 8 p.

Additional Information Sources

American Railway Engineering Association. 1986-87. The manual of recommended practice for railway engineering and maintenance of right-a-way: Chapter 3, Ties and Wood Preservation. American Railway Engineering Association, 50 F Street, N.W., Suite 7702, Washington, DC 20001.

ASTM. 1986. Annual Book of ASTM Standards, Part 22 Wood: Adhesives. American Society for Testing and Materials, 1916 Race St., Philadelphia, PA 19103.

AWPI. 1993. Answers to often-asked questions about treated wood. American Wood Preservers Institute, 1945 Old Gallows Road, Suite 550, Vienna, VA 22181.

Baker, A.J. 1980. Durability of building materials and components: Corrosion of metal. ASTM STP 691. Sereda, P.J.; Litvan, G.G. eds., Philadelphia, PA: American Society for Testing and Materials: 981-993.

Chudnoff, Martin. 1984. Tropical timbers of the world. Agric. Handb. 607. Washington, DC: U.S. Department of Agriculture. 464 p.

Comfort, M. 1993. Environmental and occupational health aspects of using CCA treated timber for walking track construction in the Tasmanian Wilderness World Heritage Area. Tasmanian Parks and Wildlife Service, Department of Environment and Land Management, Hobart, Tasmania, 7001. 41 p.

Curtis, A.B. Jr.; Williams, Lonnie H. 1990. Borates offer effective protection with less hazard to the environment. USDA Forest Service, Cooperative Forestry Utilization Report R8-UR6. 4 p.

Environmental considerations in the manufacture, use, and disposal of preservative-treated wood. 1993. Forest Products Society, Madison, WI. 109 p.

Eslyn, Wallace E.; Clark, Joe W. 1979. Wood bridges—decay, inspection and control. Agric. Handb. 557. Washington, DC: U.S. Department of Agriculture. 32 p.

Extension Service. [n.d.]. Apply pesticides correctly—A guide for private and commercial applicators. [n.d.] Extension Service, U.S. Department of Agriculture and the U.S. Environmental Protection Agency. 128 p.

Gjovik, Lee R.; Baechler, Roy H. 1977. Selection, production, procurement and use of preservative-treated wood. Supplementing Federal Specification TT-W-571. Gen. Tech. Rep. FPL-GTR-15. Madison, WI: U.S. Department of Agriculture Forest Service, Forest Products Laboratory. 37 p.

Gjovik, L.R.; Davidson, H.L. 1975. Comparison of wood preservatives in Mississippi post study. Res. Note FPL-RN-01. Madison, WI: U.S. Department of Agriculture, Forest Service, Forest Products Laboratory. 15 p.

Gjovik, Lee R.; Schumann, David R. 1992. Treatability of native softwood species of the northeastern United States. Res. Pap. FPL-RP-508. Madison, WI: U.S. Department of Agriculture Forest Service, Forest Products Laboratory. 20 p.

Graham, R.D. [n.d.] Wood preservation minicourse (slide-tape package). Forestry Media Center, Oregon State University, Corvallis, OR.

Graham, Robert D.; Helsing, Guy G. 1979. Wood pole maintenance manual: inspection and supplemental treatment of Douglas-fir and western red cedar poles. Res. Bull. 24. Corvallis, OR: Oregon State University, Forest Research Laboratory. 62 p.

Gutzmer, D.I., comp. 1991. Comparison of wood preservatives in stake tests. Res. Note FPL-RN-02. Madison, WI: U.S. Department of Agriculture Forest Service, Forest Products Laboratory. 123 p.

Hamel, Margaret, ed. 1990. First International Conference of Wood Protection with Diffusible Preservatives. *In*: Proceedings 47355. Madison, WI: Forest Products Research Society. 143 p.

Helsing, Guy G. 1979. Recognizing and controlling marine wood borers. Publication 49. Oregon State University Extension Service, Corvallis, OR.

Helsing, Guy G.; Morrell, Jeffrey J.; Graham, Robert D. 1986. Service life of Douglas-fir piles: Methods for protecting cutoff tops from decay. Forest Products Journal. 36(2): 21-24.

Highley, Terry L.; Eslyn, Wallace E. 1982. Using fumigants to control interior decay in waterfront timbers. Forest Products Journal. 32(2): 32-34.

Johnson, Bruce R. 1986. Protection of timber bulkhead from marine borers. Geotech Spec. Pub. 7. New York: American Society of Civil Engineers: 16-34.

Johnson, Bruce R. 1990. Comparison of preservative treatments in marine exposure of small wood panels. Res. Note FPL-RN-0258. Madison, WI: U.S. Department of Agriculture Forest Service, Forest Products Laboratory. 28 p.

Keeney, C.A.; Pollio, S.E. 1984. Evaluation of nondestructive underwater timber inspection techniques. Tech. Note N-1703. Port Hueneme, CA: Naval Civil Engineering Laboratory. 58 p.

Koch, Peter. 1972. Utilization of the southern pines. Vol. 1 and 2. Agric. Handb. 420. Washington, DC: U.S. Department of Agriculture. 1,663 p.

Kollman, Franz F.P.; Cote, Wilfred A. Jr. 1968. Principles of wood science and technology. Vol. 1 solid wood. Springer-Verlag. 592 p.

Levi, Michael; Machemehl, Jerry, eds. 1977. Wood in marine structures. UNC Sea Grant Pub. UNC-SG-77-12. *In*: Proceedings of a seminar. Sept. 1977. 92 p.

Levy, Michael P. [n.d.] A guide to the inspection of existing homes for wood-inhabiting fungi and insects. Washington, DC: U.S. Department of Housing and Urban Development, Office of Policy Development and Research. 104 p.

Marer, Patrick J.; Grimes, Mark. 1992. Wood preservation. Publication 3335. University of California, Statewide Integrated Pest Management Project. Division of Agricultural and Natural Resources. 86 p.

Meyer, Robert M.; Kellogg, Robert M., eds. 1982. Structural use of wood in adverse environments. New York: Van Nostrand Reinhold Co. 510 p.

Moore, Harry B. 1979. Wood-inhabiting insects in houses: Their identification, biology, prevention and control. U.S. Department of Agriculture, Forest Service, and Department of Housing and Urban Development. 133 p.

Morrell, Jeffrey J.; Corden, Malcolm E. 1986. Controlling wood deterioration with fumigants: A review. Forest Products Journal. 36(20): 26-34.

Morrell, Jeffrey; Helsing, Guy G.; Graham, Robert D. 1984. Marine wood maintenance manua[illegible] A guide for proper use of Douglas-fir in marine exposures. Res. Bull. 48. Corvallis, OR: Forest Research Laboratory, Oregon State University. 62 p.

Nicholas, Darrel D., ed. 1973. Wood deterioration and its prevention by preservative treatment. Vol. 1 and 2. Syracuse, NY: Syracuse University Press. 380 and 402 p.

Panshin, A.J.; De Zeeuw, Carl. 1980. Textbook of wood technology. 4th ed. New York: McGraw-Hill Book Company. 722 p.

Scheffer, T.C. 1991. Damage to west coast wood structures by decay fungi, insects and marine [illegible]ers. Special Pub. 22. Corvallis, OR: College of F[illegible]try, Forest Research Laboratory, Oregon State Uni[illegible]y. 32 p.

Scheffer, T.C.; Verrall, A.F. 1973. Principles for pro[illegible]cting wood buildings from decay. Res. Pap. FPL-R[illegible]-190. Madison, WI: U.S. Department of Agriculture Forest Service, Forest Products Laboratory. 55 p.

SFPA. 1992. A guide to Southern Pine shakes. Southern Forest Products Association, Box 641700, Ke[illegible]er, LA 70064. 7 p.

Stalker, Ian N.; Applefield, Milton. 1986. The preservation of wood. A self study program for wood treaters. Athens, GA: Cooperative Extension Service, University of Georgia. 128 p.

Thompson, Warren S.; Koch, Peter, 1981. Preservative treatment of hardwoods: A review. Gen. Tech. Rep. SO-35. New Orleans, LA: U.S. Department of Agriculture Forest Service, Southern Forest Experiment Station. 47 p.

Winandy, J.E. 1994. Impact of preservative and fire-retardant treatments on allowable design stresses for wood. Wood Design Focus. 2(1): 14-16.

WWPI. 1990. Pressure treated wood for marine, commercial, industrial and highway construction. Western Wood Preservers Institute, 601 Main St., Suite 401, Vancouver, WA 98660. 8 p.

WWPI. 1993. Guide to the characteristics, use and specifications of pressure-treated wood. Western Wood Preservers Institute, 601 Main St., Suite 401, Vancouver, WA 98660. 8 p.